RFID for Libraries

CHANDOS
INFORMATION PROFESSIONAL SERIES

Series Editor: Ruth Rikowski
(email: Rikowskigr@aol.com)

Chandos' new series of books is aimed at the busy information professional. They have been specially commissioned to provide the reader with an authoritative view of current thinking. They are designed to provide easy-to-read and (most importantly) practical coverage of topics that are of interest to librarians and other information professionals. If you would like a full listing of current and forthcoming titles, please visit our website www.chandospublishing.com or email info@chandospublishing.com or telephone +44 (0) 1223 891358.

New authors: we are always pleased to receive ideas for new titles; if you would like to write a book for Chandos, please contact Dr Glyn Jones on email gjones@chandospublishing.com or telephone number +44 (0) 1993 848726.

Bulk orders: some organisations buy a number of copies of our books. If you are interested in doing this, we would be pleased to discuss a discount. Please email info@chandospublishing.com or telephone +44 (0) 1223 891358.

RFID for Libraries

A practical guide

M. Paul Pandian

Chandos Publishing
Oxford · Cambridge · New Delhi

Chandos Publishing
TBAC Business Centre
Avenue 4
Station Lane
Witney
Oxford OX28 4BN
UK
Tel: +44 (0) 1993 848726
Email: info@chandospublishing.com
www.chandospublishing.com

Chandos Publishing is an imprint of Woodhead Publishing Limited

Woodhead Publishing Limited
Abington Hall
Granta Park
Great Abington
Cambridge CB21 6AH
UK
www.woodheadpublishing.com

First published in 2010

ISBN:
978 1 84334 545 9

British Library Cataloguing-in-Publication Data.
A catalogue record for this book is available from the British Library.

Typeset by RefineCatch Limited, Bungay, Suffolk

Contents

List of abbreviations

AFI	application family identifier
AIDC	automatic identification and data capture
API	application program interface
CAC	common access card
DHCP	dynamic host configuration protocol
EAS	electronic article surveillance
EM	electro magnetic
EPC	electronic product code
GPS	geographic positioning system
HF	high frequency
IEC	International Electrotechnical Commission
ILS	integrated library system
IMSc	Institute of Mathematical Sciences
IP	internet protocol
IS	information system
ISO	International Organization for Standardization
IT	information technology
LAN	local area network
LED	light emitting diode
LF	low frequency
LMS	library management system
MARC	machine readable cataloging
MOM	message oriented middleware
NCIP	NISO circulation interchange protocol
NISO	National Information Standards Organization
NIST	National Institute of Standards and Technology
OECD	Organization for Economic Cooperation and Development
PC	personal computer
PDA	personal data assistant

RFDC	radio frequency data communication
RFID	radio frequency identification
ROI	return on investment
SIP	standards interchange protocol
TCP	transmission control protocol
UDP	user datagram protocol
UHF	ultra high frequency
URL	uniform resource locator
USB	universal serial bus
WWW	World Wide Web

List of figures and tables

Figures

Tables

About the author

M. Paul Pandian obtained his PhD from Karnatak University, Dharwad, India and an Associateship in Documentation and Information Science from the Documentation Research and Training Centre, Indian Statistical Institute, Bangalore, India. He is currently the Head of the Library and Information Resource Centre at the Institute of Mathematical Sciences, Department of Atomic Energy, Chennai, India where he has recently successfully implemented an RFID-based system for the library. He was previously the Head of the Library and Information Resource Centre at the Indian Institute of Management, Indore, India and a member of the core team that was responsible for the setting up of a campus-wide information system for the IIM. He has also worked as a scientist at the INFLIBNET Centre, University Grants Commission, India where he was responsible for developing the online union catalogs of participating libraries at INFLIBNET. As a course coordinator at INFLIBNET, he also designed and developed course materials for a six-week residential course on the applications of computer and communication technologies in libraries for library executives and information scientists. He has in addition contributed several research articles on the topic of library and information science to a number of journals and presented papers at national and international conferences.

The author may be contacted at:

M. Paul Pandian
Head, Library and Information Resource Centre
Institute of Mathematical Sciences
CIT Campus, Taramani
Chennai 600 113
Tamil Nadu
India

E-mail: *pandian@imsc.res.in*

1

Introduction

In an era marked by change, uncertain economic conditions, and relentless competition, organizations are striving to attain process efficiencies that will enable them to drive down costs and provide competitive advantage. The evolution and application of technologies have always played a key role in improving operational performances. Technological advancements open the door for new applications that were not imaginable or possible before. As new technology is developed and its potential is proven, organizations contemplate using it in processes and equipment that can generate value for their customers while improving their organization's operational performance in terms of cost, quality, speed, and flexibility. Organizations are also applying advanced technologies to strengthen managerial ability to enhance organizational change and growth for better operations (Apte et al., 2006).

Organizations utilize modern information systems (IS) to acquire, interpret, retain, and distribute information. Innovations in information technology (IT) continue to improve the cost-performance capabilities of organizations to perform these four basic IS tasks. For example, the Internet has dramatically altered the capability of the firm to acquire external data and distribute it throughout and beyond the organization. Intelligent agents and knowledge management systems allow managers to interpret data and information to create useful managerial knowledge. Technical improvements in storage media allow firms to amass vast data warehouses, while ever increasing processing power allows managers to mine their data for useful information about their operations, existing customers, and potential markets. Further, advances in technology-based real-time information gathering and decision support systems promote real-time decision making that allow firms to refine operational performance (Curtin et al., 2007).

Throughout history, there has been a need to identify 'things.' By identifying things, we can sort, classify, request, ship, account for, and look for specific objects. We can do so for our personal use, for business purposes, and even for governmental functions. As a society, we have come to expect that certain 'things' would be – must be – uniquely identified. Today, we are uniquely identified by a variety of entities, including (Wyld, 2005):

- by the government, through social security numbers;
- by employers, through employee ID numbers;
- by universities, through student ID numbers;
- by insurers, banks, credit card companies, and other financial institutions, through account numbers.

While we have seen it is historically necessary to uniquely identify such highly important assets as ourselves, the vast majority of 'things' have remained identified by their class, category, or type. Until two decades ago, the human eye served as the primary mechanism for discriminating between objects of different types, whether they are different species of trees, different brands of ketchup, or different forms of munitions. However, with the advent of barcode technology, for the first time, machines – in addition to people – could identify objects (Wyld, 2005).

Ever since barcode became the dominant standard in the last century, there were many theorists and practitioners who realized that there are great limitations to its use and further development. These people were looking for something else: new technology, a new approach, something that will be able to satisfy the ever increasing variety of demand for 'next generation barcode.' While many were searching for the answer in the new 'space age' technology, others realized that the technology was already there, in radio waves. Using radio waves was in many ways superior to what barcode was able to provide to its users. The good abilities of radio waves and their attributes were well known; so they had numerous applications such as radio broadcasting, wireless telegraphy, telephone transmission, television, radar, navigational systems, and space communication (Bumbak, 2005).

However, many modern technologies give the impression that they work by magic, particularly when they operate automatically and their mechanisms are invisible. A technology called radio frequency identification (RFID), which is relatively new to the mass market, has exactly this characteristic and seems a lot like magic to many people. RFID is an electronic tagging technology that allows an object, place, or person to be automatically identified at a distance without a direct line-of-sight, using electromagnetic waves (Want, 2004).

The term 'RFID' has become a general term used to describe sensory technology that uses radio waves to scan and identify separate and distinct items. RFID is only one of numerous technologies grouped under automatic identification and data capture (AIDC) technologies, such as barcode, magnetic inks, optical character recognition, voice recognition, touch memory, smart cards, biometrics, etc. AIDC technologies have been used for decades to increase accuracy and efficiency in the data collection process for many activities. At their core, all AIDC technologies support two common goals:

- to eliminate errors associated with identification and/or data collection, and
- to accelerate the throughput process.

RFID is a representative technology of AIDC technologies. Barcoding and RFID provide quick, more accurate, and cost-effective ways to identify, track, acquire and manage data and information about items, personnel, transactions, and resources. With RFID technology, we have the advantages of faster and multiple ID recognition, easy to use operational interface, etc., compared to the barcode system, which has been the dominating AIDC technology (Chao et al., 2005).

Although RFID technology has been around for a long time, it has only had a surge in its acceptance and a massive growth in its use in the last few years. Several developments have sped up the adoption of the technology:

- First, technical standards are being established.
- Second, the cost of the tags has come down.
- Third, mandates to use RFID are being issued by major retailers and organizations.

The use of RFID technology creates opportunities in all realms of life – for business, science, government, and leisure-time activities. RFID can optimize processes, facilitate traceability, guarantee authenticity, improve product safety, boost efficiency, and simplify access control. RFID has recently emerged as one of the emerging technologies for asset tracking, inventory management, supply chain management, payment systems, information sharing, access control, and security using radio waves. Industries with the greatest opportunities to use RFID include retail, aerospace, defense, health care, logistics, pharmaceutical organizations, and libraries.

What is RFID?

Technology, over time, allows for the improvement and creation of better products and devices. Often the new technology is phased in over long periods of time while still being refined. However, on occasion there is a new technology so potent that it is implemented in a flurry, and so quietly and pervasively that the typical user may be using it without understanding how it works or what are its possible implications. Perhaps there is no better example of such a situation than with the current and future utilization of RFID, which will soon exist in every object created. RFID is a rapidly emerging technology that will surely have a dramatic global impact on how goods are exchanged and how authenticity and security are provided, so much to the point that the technology will become integrated into people's daily lives and help drive business for the next few decades (Gragg, 2003).

At first glance, the concept of RFID and its application seems simple and straightforward. But in reality, the contrary is true. RFID is a technology that spans systems engineering, software development, circuit theory, antenna theory, radio propagation, microwave techniques, receiver design, integrated circuit design, encryption, materials technology, mechanical design, and network engineering, to mention a few.

RFID stands for *radio frequency identification*. Considered as a wireless AIDC technology, RFID refers not only to the tag containing a chip but also to an antenna for sending and receiving data, an interrogator, also called reader, and its antennas to communicate through radio frequency with the tag, and finally, a middleware that manages, filters, aggregates and routes the data captured. All these elements are essential to constitute a 'basic' RFID system (Asif and Mandviwalla, 2005).

RFID has been around for some 50 years, but lack of relevant technological knowledge prevented its development. Now, thanks to recent achievements in information and communication technologies, RFID can be used in many more situations, particularly in business processes (Smith, 2004). RFID technology builds a bridge between the physical world of a product and the virtual world of digital data (Heng, 2004). Although it is often thought that RFID and barcodes are competitive technologies, they are in fact complementary in some aspects. RFID helps overcome some of the drawbacks associated with barcode technology. Barcodes have one significant downfall – they require line-of-sight technology. That means the scanner has to see the barcode to read it, which usually means items have to be manually oriented towards the scanner for it to be read. Compared to barcodes, RFID tags are 'smarter:' the information on the microchip can be read automatically, at a distance, by another wireless machine. This means RFID is easier to use and more efficient than barcodes: there is no need to pass each individual object/animal/person in front of a scanner to retrieve the information contained in each tag. Following are significant advantages to using RFID tags:

- RFID tags can be read rapidly in bulk to provide a nearly simultaneous reading of contents, such as items in a stockroom or in a container.
- RFID tags can be read in no-line-of-sight conditions (e.g. inside packaging or pallet).
- RFID tags are more durable than barcodes and can withstand chemical and heat environments that would

destroy traditional barcode labels. Barcode technology does not work if the label is damaged.

- RFID tags can potentially contain a greater amount of data compared to barcodes, which commonly contain only static information such as the manufacturer and product identification. Therefore tags can be used to uniquely identify an object.
- RFID tags do not require any human intervention for data transmission.

RFID tags can be placed on all kinds of objects such as consumer goods, shipping containers, high-value equipment, and even human beings so that their movement and location can be easily tracked. RFID systems also can be linked with video security systems. Linking video and access control are good ideas for night applications. A camera can pan-tilt-zoom and also link the access control transaction history with camera data to look at events triggered in the systems.

History of RFID

It is difficult to trace the history of RFID technology back to a well-defined starting point as there is no clear progression of RFID developments over time that ultimately arrives at the present state of matters. The origins of RFID technology lie in the nineteenth century when luminaries of that era made great scientific advances in electromagnetism. Of particular relevance to RFID are: Michael Faraday's discovery of electronic inductance, James Clerk Maxwell's formulation of equations describing electromagnetism, and Heinrich Rudolf Hertz's experiments validating Faraday and Maxwell's predictions. Their discoveries laid the foundation for modern radio communications. The history of RFID technology is intertwined with that of many other communications technologies developed throughout the twentieth century. These technologies include computers, information technology, mobile phones, wireless LANs, satellite communications, GPS, etc. Research and advances in the following three areas have given rise to commercially viable RFID (Michalek and Vaculik, 2008):

- *Radio frequency electronics.* Research in this field, as applied to RFID, was begun during World War II and continued through the 1970s. The antenna systems and RF electronics employed by RFID interrogators and tags have been made possible because of radio frequency electronic research and development.
- *Information technology.* Research in this field began in the mid-1970s and continued through to roughly the mid-1990s. Both the host computer and the interrogator/reader employ this technology. The networking of RFID readers has also been made possible by research in this area.
- *Materials science.* Breakthroughs in materials science technology in the 1990s finally made RFID tags cheap to manufacture and this has overcome cost barriers to make RFID technology commercially viable.

RFID had been hyped as a revolutionary new technology in recent years. But like the Internet, RFID has a long history stretching back to early applications in the military. The US and UK governments used an early form of RFID on airplanes in World War II to determine if the plane was a friend or a foe. Perhaps the first work exploring RFID is the landmark 1948 paper by Harry Stockman, entitled 'Communication by Means of Reflected Power' (Stockman, 1948). Stockman predicted that '... considerable research and development work has to be done before the remaining basic problems in reflected-power communication are solved, and before the field of useful applications is explored.' It required thirty years of advances in many different fields before RFID became a reality. Table 1.1 illustrates the history of RFID since 1940 (Landt, 2001).

RFID adoption

Technology innovation is widely recognized as an important driver of business transformation and economic growth. History tells us that the path to acceptance within the business community can be long for technological innovations. For example, the Internet has its origins in the late 1960s and 1970s, and did not reach wide acceptance until the late 1990s. The primary catalyst for widespread adoption came with a change in the business perceptions

Table 1.1 History of RFID

Decade	Event
1941–1950	Radar refined and used, major World War II development effort. RFID invented in 1948.
1951–1960	Early explorations of RFID technology, laboratory experiments.
1961–1970	Development of the theory of RFID. Start of applications field trials.
1971–1980	Explosion of RFID development. Tests of RFID accelerate. Very early adopter implementations of RFID.
1981–1990	Commercial applications of RFID enter mainstream.
1991–present	Emergence of standards. RFID widely deployed. RFID becomes a part of everyday life.

of value based on the advent of fast, reliable, and low-cost hypertext markup language applications. In other words, the perceived benefits or risks that are held by the users of each technological innovation influence the rate of acceptance (Venkatesh et al., 2003).

Rogers (2003) introduces five factors that accelerate or slow down the adoption and diffusion of innovations:

- relative advantage;
- complexity;
- trialability;
- observeability;
- compatibility.

Relative advantage expresses the degree to which a new technology or innovation is perceived as being superior to that currently used. The degree of relative advantage is often described as economic profitability (e.g. by reducing costs), as conveying social prestige (e.g. status seeking/motivation), or in other ways. The higher the relative advantage of an innovation is perceived by members of a social system, the more likely is its adoption. The complexity of an innovation is the degree to which it is perceived as relatively difficult to understand, to implement in an existing infrastructure, and to use. The complexity of an innovation, as perceived by members of a social system, is negatively related to its rate of adoption. Trialability is the degree to which an innovation and especially a new technology may be experimented with on a limited basis. For example, RFID pilot studies provide first experiences if the technology is able to work under a company's conditions and in a certain environment. The trialability of an innovation, as perceived by the members of a social system, is also positively related to its rate of adoption. Observeability is the degree to which the results or increases in efficiency of an innovation are visible to others. For example, a decrease of stock-outs is more visible than an improvement of a single piece of hardware of a computer system. Therefore, the observeability of an innovation, as perceived by the members of a social system, is positively related to its rate of adoption. Compatibility is the degree to which an innovation is experienced as consistent with the needs, past experiences, and existing values of potential adopters. Additionally, compatibility describes the fit and ease of integration of a technology into an existing (IT) infrastructure and is consequently positively related to the rate of adoption.

The progress of RFID adoption divides naturally into following eras (Glover and Bhatt, 2006):

- the Proprietary era;
- the Compliance era;
- the RFID-Enabled Enterprise era;
- the RFID-Enabled Industries era; and
- the Internet of Things era.

In the beginning, during the Proprietary era, businesses and governmental entities created systems designed to track one particular type of item, and this tracking information typically remained within the same business or governmental entity. In the Compliance era (the present era), businesses implement RFID to meet mandates for interoperability with important customers or regulatory agencies but often don't use the RFID data themselves. The future will bring the era of the RFID-Enabled Enterprise, where organizations will use RFID information to improve their own processes. The era of RFID-Enabled Industries will see RFID information shared among partners over robust and secure networks according to well-established standards. The final RFID era that is currently foreseeable is the era of the Internet of

Things. By this time, the ubiquity of RFID technology and other enabling technologies, combined with high standards and customer demand for unique products based on this infrastructure, will lead to a revolutionary change in the way we perceive the relationship between information and physical objects and locations. More and more, we will expect most objects in our daily lives to exist both in a particular place, with particular properties, and in the information spaces we inhabit.

RFID standards

The need for standards has become apparent to almost every one. Standardization is an important aspect of any technology from its incubation through to maturity. Standardization is advantageous for both vendors and customers. For vendors, the benefits of standardization include:

- market broadening and global competition;
- products and applications interoperability;
- cost reduction;
- fast technology acceptance and technology advancement.

For customers, standards help to achieve the following objectives:

- increase confidence in a new technology;
- facilitate applications development and reduce customization;
- reduce cost for equipment and software;
- increase decision flexibility;
- allow development of non-proprietary solutions that are not confined within one customer's organization.

Any technology is best applied in the industry when it is generally available for multiple sources and interoperability. The purpose of standardization is to define the most efficient platform on which an industry can operate and advance. Poor standardization might cancel any benefit of the technology. Two kinds of standards affect RFID:

1. Hardware or technology standards address equipment issues.
2. Software or application standards address the arrangement and handling of the data that is handled by the equipment. RFID readers and tags must conform to the same standards and designs to be interoperable. These standards and designs also can be used to coordinate the use of certain tags across multiple enterprises and in the supply chain. Common standards and designs may facilitate training, future equipment procurement, and equipment upgrades. Some readers and some tags can operate using multiple standards.

One of the factors propelling RFID technology to prominence today is the fact that the RFID industry through the extraordinary collaboration of vendors, retailers, manufacturers, governments, and academic institutions is forging a set of standards for RFID technology. Several RFID standards exist and their applications are under debate within the RFID development community. These standards cover (Weinstein, 2005):

- identification, the coding of unique item identifiers, or other data on the RF tag;
- data and system protocols, effectively the middleware of an RFID system;
- the air interface, that is, the wireless communication between the reader and the tag;

- application support, which provides advice about how to implement the technology;
- testing, compliance, and health and safety, that is, the rules that govern RFID operations; and
- terminology.

There are three major advantages of developing international standards for RFID systems. First of all, a common RFID standard will ensure interoperability among tags and readers manufactured by different vendors and allow for seamless interoperation across national boundaries. Second, due to compatibility and exchangeability, the demand for RFID components and equipment will be high, and that can bring the cost down. Finally, an internationally accepted RFID standard will facilitate the growth of the worldwide RFID market.

Standardization is expected to go hand-in-hand with market adoption. If prices continue to drop, technology continues to improve and applications continue to increase, the electronic security industry could find many more places to utilize RFID technology in the future.

Two organizations are mostly involved in drafting standards for RFID technology: the International Organization for Standardization (ISO) and EPCglobal (electronic product code). The ISO is responsible for a variety of standards, regulating air interfaces, data protocols, and applications, for instance. The standard developed by EPCglobal is directed primarily at trade, allocating different products a unique code similar to barcode. There are separate standards for contactless smart cards and for item management. The following are the most popular RFID standards (Karygiannis et al., 2007):

- ISO/IEC 14443 describes proximity smart cards that have an intermediate range up to 10cm and operate at 13.56 MHz. The standard contains four parts:

1. physical characteristics;
2. radio frequency power and signaling;
3. initialization and anti-collision; and
4. transmission protocols.

ISO/IEC 14443 has two variants known as ISO/IEC 14443A and ISO/IEC 14443B, which have different communications interfaces. Readers who are ISO/IEC 14443 compliant must be able to communicate using ISO/IEC 14443A and ISO/IEC 14443B. ISO/IEC 14443A parts 1 through 4 are used in the DoD Common Access Card (CAC), which serves as an identification card.

- ISO/IEC 15693 operates at 13.56 MHz and describes vicinity smart cards that can be read from a further distance than proximity cards. Such cards have a range of up to approximately 1 meter. The ISO 15693 specification has three main parts:
  - physical characteristics;
  - signal interface; and
  - transmission protocol.

  It holds the promise of interoperability (at the technical level) among different suppliers of RFID solutions.

  ISO 15693 is not to be confused with ISO 15963 that is used for RFID for item management – unique identification of RF tag.
- ISO/IEC 18000 is an RFID standard for item management and describes the air interface for various frequencies. Each standard within the ISO/IEC 18000 family defines communication parameters and applies to a specific electromagnetic frequency (Karygiannis et al., 2007):
  - ISO/IEC 18000-1 covers general parameters and ISO/IEC 18000-2 through 18000-7 cover specifics for particular frequency ranges.

- ISO/IEC 18000-2 covers frequencies below 135 kHz. It has two types, A (full duplex) and B (half duplex). These types are different on the physical layer. A full duplex tag can communicate with a reader while the reader is simultaneously communicating with the tag. A half duplex tag supports bi-directional communication with a reader but only one device, the tag or the reader, can communicate at a time.
- ISO/IEC 18000-3 covers frequencies operating at 13.56 MHz and describes two non-interfering and not interoperable modes of operation. Both modes use a 64-bit identifier:
  - Mode 1 has a locking feature that is not protected by a password. If the tag receives the lock command, it locks the corresponding area of memory permanently. Lock can be applied selectively to different blocks of memory.
  - Mode 2 has a 48-bit password used to protect memory access. The tag can be configured to require or not require this password. If required, then read and write commands will require the reader to issue the correct 48-bit password. The lock command can be used to permanently write protect a block of memory. Mode 2 also has a 16-bit lock pointer that is located in unaddressable memory. The lock pointer points to a word in memory. All complete blocks of memory at addresses less than the number stored in the lock pointer cannot be overwritten.
- ISO/IEC 18000-4 covers systems operating at 2.45 GHz. This standard has two modes: a passive tag reader-talks-first mode and a battery assisted tag-talks-first mode.
- ISO/IEC 18000-5 was developed for 5.8 GHz operation but this standard was withdrawn.

- ISO/IEC 18000-6 defines three types of tags. Types A and B operate at 860–930 MHz, but they use different encoding and anti-collision methods on the forward channel. Type C is equivalent to the EPCglobal Class-1 Generation-2 standard.
- ISO/IEC 18000-7 is an RTF protocol for an RFID system that operates at 433 MHz. Tags have a 32-bit tag ID and a 16-bit manufacturer ID. Readers are given a 16-bit identifier as well. A 32-bit password can be set on the tags. A bit, referred to as the 'secure bit' in the standard, is set to determine if the tag is password protected or not. If protected, read/write of the user ID, user ID length, routing code, and memory are password protected. ISO/IEC 18000-7 supports optional command database query commands that are transmitted to all tags. The queries are sent in multiple steps and can use logical operators such as clear, and, and/or, and relational operators such as equal, less than, greater than, and not. Tags that receive all steps of the query will do an internal database search and readers can retrieve the results of these queries.

The current standard used in libraries ISO 15693 was not designed for the item-level tracking done in libraries. Yet, most library RFID tags follow this standard. ISO 15693 was designed for supply chain applications. It defines the physical characteristics, air interface, and communication protocol for RFID cards (Molnar and Wagner, 2004; R. Moroz Ltd., 2004).

Library RFID applications must be able to integrate with the library's integrated library systems (ILS). The two most popular protocols that facilitate the smooth integration of RFID products and ILS are SIP2 (Standard Interchange Protocol 2) and NCIP (NISO Circulation Interchange

Protocol). Standard Interchange Protocol (SIP) was originally developed and published by 3M to allow their self-issue machines to exchange circulation messages with library systems. Features include identification of borrower and issuing material. SIP2 (version 2 of SIP) is the de-facto standard for the exchange of circulation data and transactions between different systems. It is now in use by a variety of self-issue systems, telephone renewals, PC bookings software, library security systems, and RFID systems. Over the years, shortcomings have been identified in SIP2 and the standard has been diluted as vendors attempt to modify the protocol to suit their needs. To address the shortcomings of SIP2, the National Information Standards Organization (NISO) convened a standards development group with the mission of designing a protocol that would encourage interoperability among disparate circulation, interlibrary loan, self-service, and related applications. The outcome of this group was NCIP. NCIP was approved by NISO in 2002.

ISO 28560, which was still a draft standard as of mid-2009, is essential to interoperability among RFID systems. The adoption of ISO 28560 will not guarantee interoperability as vendors may still seek to encrypt the data on their tags, add proprietary security functions, and/or include software or firmware that is system dependent and can only be used with specific tags (Boss, 2007).

RFID system components

The purpose of an RFID system is to enable data to be transmitted by a mobile device, called a tag, which is read by an RFID reader and processed according to the needs of a particular application. RFID tags and readers have to be

Figure 1.1 RFID system components

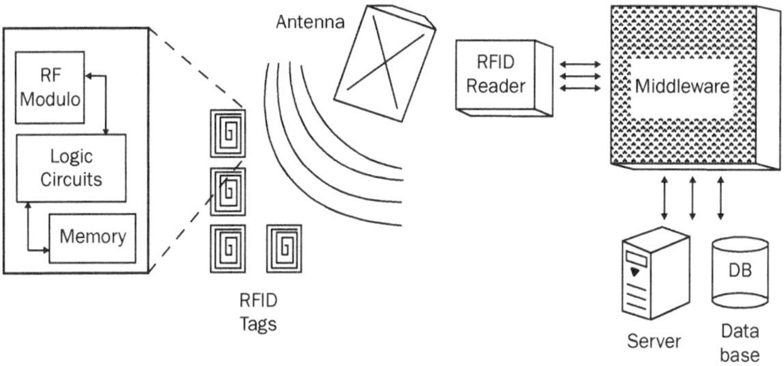

Source: Candino et al. (2009).

tuned to the same frequency to communicate. In a typical RFID system, individual objects are equipped with a small, inexpensive tag, which contains a digital memory chip that is given a unique electronic product code. The interrogator, an antenna packaged with a transceiver and decoder, emits a signal that activates the RFID tag so it can data read from and write to it. When an RFID tagged item passes through the electromagnetic zone, it detects the reader's activation signal. The reader decodes the data encoded in the tag's integrated circuit (silicon chip) and the data is passed on to the host computer for processing. Figure 1.1 illustrates the functional components of RFID system.

RFID systems may be roughly grouped into four categories (AIM):

- *Electronic article surveillance (EAS) systems*. These are typically a one bit system used to sense the presence/ absence of an item. The large use for this technology is in retail stores where each item is tagged and a large antenna readers are placed at each exit of the store to detect unauthorized removal of the item (theft).

- *Portable data capture systems.* These are characterized by the use of portable data terminals with integrated RFID readers and are used in applications where a high degree of variability in sourcing required data from tagged items may be exhibited. The hand-held readers/portable data terminals capture data, which is then either transmitted directly to a host information management system via a radio frequency data communication (RFDC) link or held for delivery by line-linkage to the host on a batch processing basis.
- *Networked systems.* These applications can generally be characterized by fixed position readers deployed within a given site and connected directly to a networked information management system. The transponders are positioned on moving or moveable items, or people, depending upon application.
- *Positioning systems.* These use transponders to facilitate automated location and navigation support for guided vehicles. Readers are positioned on the vehicles and linked to an on-board computer and RFDC link to the host information management system. The transponders are embedded in the floor of the operating environment and programmed with appropriate identification and location data. The reader antenna is usually located beneath the vehicle to allow closer proximity to the embedded transponders.

RFID systems differ very significantly from one another in their features. These differences pertain to (Heng, 2009):

- degree of freedom (open vs. closed);
- data storage (centralized vs. decentralized);
- data processing (real-time vs. batch-processing);

- physical form (e.g. earmark, ceramic bolus, glass encapsulation for implantation, nail, and smart label);
- storage capacity;
- energy supply (with/without battery);
- writability of the RFID tag (e.g. read-only; write-once-read-many; read-and-write);
- radio frequency (from low frequency [LF] to ultrahigh frequency [UHF]).

RFID systems can also be distinguished by their frequency ranges and applications as given in Table 1.2.

RFID systems may comprise the following components (Lahiri, 2005):

- *Tag*. This is a mandatory component of any RFID system.
- *Reader*. This is a mandatory component, too.
- *Reader antenna*. This is another mandatory component. Some current readers available today have built-in antennas.

Table 1.2 Different RFID frequencies and their applications

Band	Frequency	Read range	Applications
LF	100–500 kHz	Up to 20 inches	Access control, animal ID, keyless entry
HF	13.56 MHz	Up to 3 feet	Access control, smart cards, libraries, anti-theft tagging
UHF	866–956 MHz	20 feet or more	Supply chain, baggage handling, toll collection
Microwave	2.45 GHz	3–10 feet	Item tracking, toll collection

- *Controller*. This is a mandatory component. However, most of the new-generation readers have this component built in to them.
- *Sensor, actuator, and annunciator*. These optional components are needed for external input and output of the system.
- *Host and software system*. Theoretically, an RFID system can function independently without this component. Practically, an RFID system is close to worthless without this component.
- *Communication infrastructure*. This mandatory component is a collection of both wired and wireless network and serial connection infrastructure needed to connect the components together to effectively communicate with each other.

As discussion of RFID technology tends to focus mainly on tags and readers, let us look at them in detail.

RFID tag

RFID tags, also called transponders, are the heart of an RFID system because they store the information that describes the object being tracked. Specific object information is stored in the memory of tags and is accessed via the radio signal of RFID readers. RFID tags that perform the data carriage consist of a microchip with some computation and storage capabilities, and a coupling element such as an antenna coil for communication that are packaged so that they can be attached to objects (Tellkamp, 2006). There are many different types of tags they and can be classified according to two main criteria:

- *the type of memory*: read-only, write-once-read-many, or fully rewritable;

- *the source of power*: active, semi-active, semi-passive, and passive.

Following is a brief description of different types of tags (NIST, 2007):

- *Passive tags* do not possess their own battery. They depend on the energy provided by the reader both for receiving and sending data. Due to power restrictions, the operating range is lower than that for active tags, but passive tags are significantly cheaper than active tags (Figure 1.2).
- *Active tags* have their own internal battery that is used to power the microchip and to send data to the reader. They thus attain higher read ranges and can read weaker signals than passive tags. Furthermore, they are able to transmit data over long distances and face less interference problems (e.g. metal and water) than passive tags. However, active tags are more expensive than passive tags and the lifespan of batteries is limited (Figure 1.3).

Figure 1.2 RFID passive tags

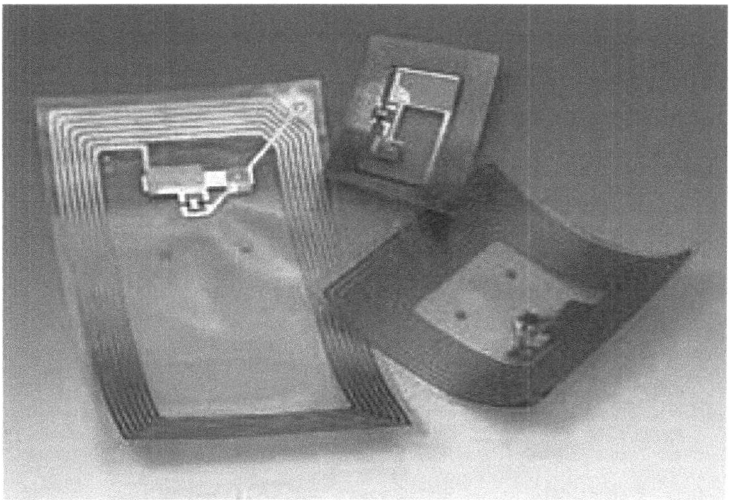

Figure 1.3 RFID active tags

- *Semi-active tags* remain dormant until they receive a signal from the reader. Therefore, they have a longer battery life than that of active tags. Similar to active tags, semi-active tags can rely on their battery to transmit data to the reader.
- *Semi-passive tags* also rely on a battery to power the microchip, but use the energy provided by the reader to transmit data. Compared to passive tags, semi-passive tags have more power for internal functions. They are for instance used to power integrated sensors.
- *Read-only tags* contain data such as a serialized tracking numbers, which are pre-written onto them by the tag manufacturer or distributor. Read-only tags are generally the least expensive, because they cannot have any additional information included as they move throughout the supply chain. Any updates to that information have to be maintained in the application software that tracks the stock unit's movement and activity.
- *Write-once tags* enable a user to write data to the tag one time during production or distribution. This information can be a serial number or other data, such as a lot or batch number.
- *Full read-write tags* allow new data to be written to the tag as needed and written over the original data.

RFID tags come in a range of shapes and sizes. The following are the most common ones:

- *label*: the tag is a flat, thin, flexible form;
- *ticket*: a flat, thin, flexible tag on paper;
- *card*: a flat, thin tag embedded in tough plastic for long life;
- *glass bead*: a small tag in a cylindrical glass bead, used for applications such as animal tagging (e.g. under the skin);
- *integrated*: the tag is integrated into the object it is tagging rather than applied as a separate label;
- *wristband*: a tag inserted into a plastic wrist strap.

Given that RFID tags will not replace barcoded labels any time soon, it is important to understand the distinct differences between these two auto-identification technologies. In comparison to RFID technology, the main advantages of the barcode include significantly cheaper tags and in the majority of cases less expensive readers. In addition to this, there are no interferences due to water or metal as it is the case for RFID. Another important point is the better control of scanned data. For example, when different barcode-labeled cases on a pallet are scanned and errors occur, it is usually obvious that cases have not been scanned correctly. However, if errors occur when RFID-tagged cases are read in bulk, it is more complicated and time consuming to detect, for example, which cases have not been read at all. However, there are also a number of significant disadvantages when compared to RFID technology. At first, barcodes are not very robust as compared to RFID tags. Poor weather conditions such as rain or extreme temperatures as well as dirt render barcode labels inoperable. In terms of data storage, storage capacity does not attain the same level as for RFID tags. Furthermore, once the particular barcode is generated, data cannot be changed or augmented and therefore it is not possible to store additional data during the production process (Hodges and McFarlane, 2005). In addition to this, it is not possible to obtain the same level of high read ranges as with RFID technology. Finally, integration of other technologies such as sensor technology is unrealizable. The disadvantages of barcodes on the technological level mirror the image of technological benefits of RFID technology. Table 1.3 summarizes these differences (Shutzberg, 2004).

RFID reader

The RFID reader, also called transceiver, is the central nervous system of the entire RFID hardware system, establishing communication with and control of this component is the most important task of any entity, which seeks integration

Table 1.3 **Difference between barcode labels and RFID tags**

Barcoded labels	RFID tags
Inexpensive (but not reusable)	Costly (though potentially reusable)
Reliable to read	Not always reliable to read
Work with virtually all products	Work with most products but have trouble with some (such as those containing metals and liquids)
Can be printed before production or printed directly on items	Must be programmed, applied, and verified individually, and data synchronization is usually required
Must be read one at a time and line of sight is required	Many tags can be read simultaneously and no line of sight is required
Written once with limited data	Can potentially be written multiple times, have higher capacity, and can be combined with sensors
Have a limited read range	Can have a longer read range

with this hardware entity. RFID readers are generally composed of an RF module, a control unit, and a coupling element to interrogate RFID tags via RF communication. Readers may have better internal storage and processing capabilities, and frequently connect to back-end databases. RFID readers are devices that convert radio waves from RFID tags into a form that can be passed to middleware software. Reader requirements vary depending on the type of task and application, and almost all applications will require multiple forms of readers to make a successful system. There are a variety of different reading systems and technologies. These include (Peris-Lopez et al., 2006):

- handheld readers that act like a handheld bar code scanner;
- RFID readers embedded into mobile data collection devices;
- fixed readers, which are mounted to read tags automatically as items pass by or near them.

The RFID readers perform a variety of functions, such as activating tags by sending querying signals, supplying power to passive tags, encoding the data signals going to the tag, and finally, decoding the data received from the tag. RFID readers communicate with tags through the method of induction known as inductive coupling. When an RFID tag passes through the field of a reader's antenna, tag's antenna detects the activation signal from the reader's antenna. The RF radiation activates the tag chip, which in turn will execute the commands from the reader; either it will write information on memory or it will transmit the information on its microchip to be picked up by the reader. The information received by the reader will be demodulated and de-codified and sent to the application database through a middleware. RFID middleware provides the interface for communication between the reader and existing databases and information management systems.

There are several characteristics of an RFID reader that determine the types of tags with which it can communicate. The most fundamental characteristic is the frequency or frequencies at which the reader's radio communicates. Readers and tags must communicate at the same frequency in order for them to couple. But, some RFID readers called dual-frequency readers can communicate at more than one frequency. Most RFID readers communicate exclusively with active tags or exclusively with passive tags. This means that an RFID reader that is manufactured to communicate with passive tags will not be able to communicate with active tags and vice versa. There may be RFID readers that can communicate with both, but, if they exist, they are not the norm. When implementing a solution that uses passive RFID tags, a reader capable of interfacing with passive tags must be employed. A key feature of an RFID reader is the number of tags that it can sample in its tag population. Some readers may be able to sample 10 tags a second while others may be able to sample 100 tags a second. The number of tags sampled per second is usually influenced by the following (Banks and Thompson, 2008):

- the anti-collision algorithm used by the tags;
- the processing capabilities of the reader that usually maps to the type and speed of processor in the reader;
- the amount of memory in the reader;
- the capabilities of the digital signal processor in the reader's radio.

The RFID reader must also communicate the tags it reads with an application that can make use of the data. The most common types of communications interfaces on this side of the reader are (Banks and Thompson, 2008):

- serial – RS232 or RS422;
- IP (Ethernet) – TCP or UDP;
- USB.

RFID readers are available in many sizes, frequencies and with different data processing and reporting capabilities. Understanding these characteristics is important for designing an RFID solution that will function properly and be maintainable. Different types of readers are given in Figures 1.4–1.7.

Figure 1.4 Active RFID reader

Figure 1.5 Active RFID Wi-Fi reader

Figure 1.6 Wi-Fi inventory reader

Figure 1.7 RFID reader – USB

A reader has the following main components (Lahiri, 2005):

- transmitter;
- receiver;
- microprocessor;
- memory;
- input/output channels for external sensors, actuators, and annunciators (although, strictly speaking, these are optional components, they are almost always provided with a commercial reader);
- controller (that may reside as an external component);
- communication interface;
- power.

Like tags, readers can also be classified using two different criteria. The first criterion is the interface that a reader provides for communication. Based on this, readers can be classified as follows (Lahiri, 2005):

- *Serial*. Serial readers use a serial communication link to communicate with an application. The reader is physically connected to a computer's serial port using an RS-232 or RS-485 serial connection. Both of these connections have an upper limit on the cable length that can be used to connect a reader to a computer. RS-485 allows a longer cable length than RS-232 does. The advantage of serial readers is that the communication link is reliable compared to network readers. Therefore, the use of these readers is recommended to minimize dependency on a communication channel. The disadvantage of serial readers is the dependence on the maximum length of cable that can be used to connect a reader to a computer. In addition, because the number of serial ports is generally limited on a host, a larger number of hosts (as compared to the number of hosts needed for network readers) might be needed to connect to all the serial readers. Another problem is maintenance – if the firmware needs to be updated, for example, maintenance personnel might have to physically deal with each reader. Also, the serial data-transmission rate is generally lower than the network data-transmission rate. These factors might result in higher maintenance costs and significant operation downtime.

- *Network*. Network readers can be connected to a computer using both wired and wireless networks. In effect, the reader behaves like a network device installation that does not require any specialized knowledge of the hardware. The advantage of network readers is that there is no dependence on the maximum length of cable that can be used to connect a reader to a computer. A smaller number of hosts are generally needed as compared to the serial readers. In addition, the reader firmware can be updated remotely over the network without any need to visit the reader physically. This can ease the maintenance effort and lower the cost of ownership of such an RFID system. The disadvantage of network readers is that the communication link is not as reliable compared to serial readers. When the communication link goes down, the back end cannot be accessed. As a result, the RFID system might come to a complete standstill. The readers, in general, have internal memory to store tag reads. This feature might somewhat alleviate short network outages.

The readers are becoming increasingly sophisticated, acting as gateways into the network centric communication systems of modern enterprises by supporting communication protocols such as TCP/IP and network technologies such as DHCP, UDP/IP, and Ethernet or 802.11x (for wirelessly sending data back to the enterprise).

RFID benefits

Five technological core benefits distinguish RFID technology from alternative technologies (OECD, 2007): traveling data storage, contactless data transmission and absence of line-of-sight, bulk reading, robustness, and the ability to integrate sensor technology. These five technological core benefits entail a variety of usage benefits. For example, highly automated data handling and authentication automated data handling requires less physical contact, reduces manual errors and allows for real-time inventory, and, generally speaking, an overall higher speed of operations and transactions. With traveling data storage, contactless data transmission and bulk reading, items can be tracked and their location can be precisely determined, which prevents shrinkage and allows for theft and diversion prevention as well as anti-counterfeiting measures. These usage benefits of RFID contribute to business benefits such as improved processes, improved inventory auditing, efficient access and exit control, and value-added services to customers.

RFID is mostly used for identifying people, objects, transactions, or events through a wireless communication connection. The development of RFID technology emerges to be one of the most interesting innovations for the improvement of business process efficiency across various sectors including the manufacturing, transportation and logistics, wholesale distribution, retail trade, and library sectors. Despite first appearing in tracking and access applications in the 1980s, the potential of RFID has only been recognized relatively recently. Using RFID tags, it is possible to identify and track objects and people without time delays, without human intervention, and thus without variable costs. With even smaller, smarter, and cheaper tags and readers, RFID is opening up amazing value chain possibilities. Through RFID technology, organizations can improve efficiency and visibility, cut costs, better utilize their assets, produce higher quality goods, reduce shrinkage, or counterfeiting and increase sales by reducing out-of-stocks. Also, RFID can gain a number of social benefits both in the private sector and the public sector. All this means that RFID will have a great impact on the processes and IT systems of organizations and public and societal organizations. Use of RFID is expected to increase the cost-effectiveness of public transport, to fight fraud, to increase social safety, and to introduce additional services. RFID offers many benefits to a wide variety of industries. In healthcare, RFID is expected to combat counterfeiting, to increase the quality of care, to improve the availability of health information, to prevent surgical mistakes, and to reduce theft of medical equipment. In the retail sector, RFID is expected to lead to less out-of-stock items, to more efficient logistics, to better consumer profiles, and to better quality information. These are clear benefits, which show up in the short return on investment periods, indicating the high economic value of the RFID implementations (Lieshout et al., 2007).

The transportation and automotive sectors have also made headway. With the backing of major global brands and increased convergence around global technical standards, RFID is gaining momentum. It can help stakeholders to reduce shrinkage, reduce material handling costs, increase data accuracy, enable supply chain business process innovation, and improve information sharing. RFID's potential benefits are large and many novel applications will emerge in the future – some of which cannot be imaginable now. Table 1.4 (Das, 2006) shows that the larger application of RFID might generate a number of socio-economic benefits. It is obvious that the price–development of the tags

Table 1.4 **Potential benefits of RFID applications in various application areas**

Application area	Potential benefits					
	Cost reduction	Increased sales	Crime reduction	Better service	Safety	Removal of tedious procedures
Library books, DVDs, etc.	Yes	–	Yes	Yes	–	Yes
Parts for aircraft and other machinery	Yes	–	Yes	Yes	Yes	Yes
Blood bags and samples	Yes	–	–	Yes	Yes	Yes
Military	Yes	–	Yes	Yes	Yes	Yes
Book retail	Yes	Yes	Yes	Yes		Yes
Drugs prescription	Yes	Yes	Yes	Yes	Yes	Yes
Postal	Yes	–	Yes	Yes	Yes	Yes
Other consumer packaged goods (CPG)	Yes	Yes	Yes	Yes	Yes	Yes

Source: Das (2006).

is only one factor in the adoption and broad application of RFID. The development of some potential markets might not be as price-sensitive as often is believed, because of the social benefits that (also) might be realized.

RFID applications

A wide variety of organizations are using RFID to fine-tune their operations. RFID is being used in an ever-increasing number of industries for such purposes as access control, package tracking, inventory management, e-government, baggage handling, fraud prevention, and even school attendance. Broadly speaking, these applications may be classified according to the major purpose of deploying RFID (Sabbaghi and Vaidyanathan, 2008):

- identification;
- authentication;
- location;
- automatic data acquisition.

Authentication applications include smart cards and automatic payments. In the ADA applications, objects such as produced items, cases, and pallets are tracked automatically and the captured data is used to derive enterprise applications such as supply chain management systems, customer relation management systems, and enterprise resource planning systems (Asif and Mandviwalla, 2005).

In general, there are two main areas of application, defined broadly as *proximity* or *short range*, and *vicinity* or *long range*. Long-range or vicinity applications can be described as track and trace applications, but the technology provides additional functionality and benefits for product authentication. Typical end-uses include, but are not limited to supply chain management, parcel and post, garment tags, library and rental sectors, and baggage tagging. Short-range or proximity applications are typically access control applications and mass transit ticketing. RFID technology applications grow rapidly and have received considerable worldwide attention as the costs of the RFID tag continually falls. The application of RFID changes gradually from people identification to product identification. The evolution of applications of RFID is shown in Figure 1.8.

RFID as an emerging technology has been successfully applied in supply chain management, manufacturing, and logistics, but its range of application extends far beyond these areas. Potential applications for RFID may be identified in virtually every sector of industry, commerce, and services where data is to be collected. The attributes of RFID are complimentary to other data capture technologies

Figure 1.8 Evolution of RFID applications

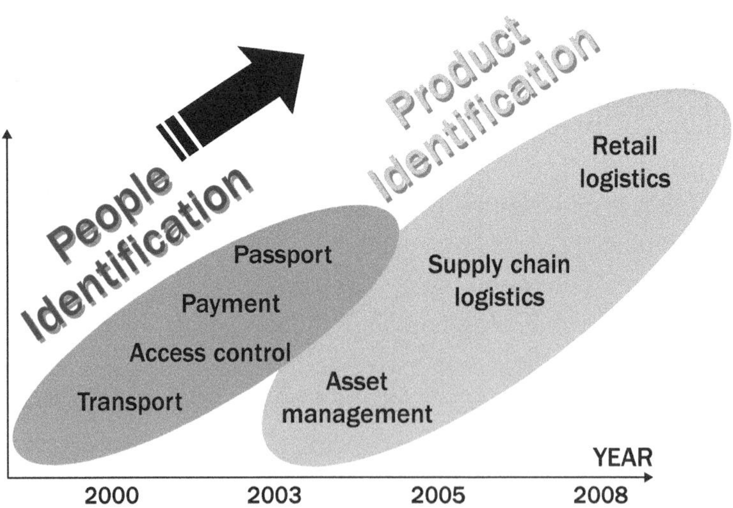

Source: Xang et al. (2006).

and thus able to satisfy particular application requirements that cannot be adequately accommodated by alternative technologies. Principal areas of application for RFID that were earlier identified included transportation and logistics, manufacturing and processing, and security (AIM).

There is tremendous potential for applying it even more widely, and increasing number of organizations have already started up pilot schemes or successfully used it in real-world environments. As standards emerge, technology develops still further, and costs reduce considerable growth in terms of application numbers, and new areas of application may be expected. Some of the more prominent specific applications include (Ngai et al., 2008):

- animal detection;
- aviation;
- building management;
- construction;
- enterprise feedback control;
- fabric and clothing;
- food safety warranties;
- health;
- library services;
- logistics and supply chain management;
- mining;
- municipal solid waste management;
- museums;
- retailing.

RFID markets

The current RFID market remains driven by high volume applications such as security/access control, contactless payment, ticketing, library management, anti-counterfeiting/authentication, ID documents (passports, national ID, etc.), item-level tagging, and animal identification (Liard and Carlaw, 2009).

The global industry for RFID technology has been steadily growing for the past few years, and is expected to pick up pace before stabilizing and settling on a steady growth path. Besides, the market is greatly benefitting from the excitement related to mandates, entry of new players, re-positioning of some organizations as RFID-focused, new product and service launches, technology advances, standards evolution, partnerships and alliances, mergers and acquisitions, and general market awareness (Liard and Carlaw, 2009).

According to the new research report, 'Global RFID Market Analysis till 2010,' the RFID market is set to grow at a rate of around 28 per cent in the period 2010–13. It is found that the current economic crisis across the world will not stall the global RFID market, but the market growth is expected to slowdown to around one third in 2009 as compared to the growth reported in previous years. It is also found that Asia-Pacific will witness the highest growth in RFID revenue owing to the rapid adoption of RFID applications in several countries, including China, India, South Korea, Taiwan, and Thailand (RNCOS, 2009).

In 2009, IDTechEx found that the value of the entire RFID market would be $5.56 billion, up from $5.25 billion in 2008. This includes tags, readers, and software/services for RFID cards, labels, fobs, and all other form factors. The majority of this spend would be on RFID cards and their associated services – totaling $2.99 billion. The market for RFID is growing and a large amount of this value is due to government-led RFID schemes, such as those for transportation, national ID (contactless cards and passports), military, and animal tagging. In total, 2.35 billion tags would be sold in 2009 versus 1.97 billion in 2008; 1.74 billion in 2006; and 1.02 billion in 2005 (Das and Harrop, 2009). The IDTechEx report provides some more data as depicted in Figure 1.9.

Conclusion

Technological advancements are the high-octane fuel that powers the continued acceptance and growth of new technologies. These advancements can provide the following advantages:

- make existing applications easier to use;
- offer more functionality;
- drive deployment costs down.

Figure 1.9 RFID markets

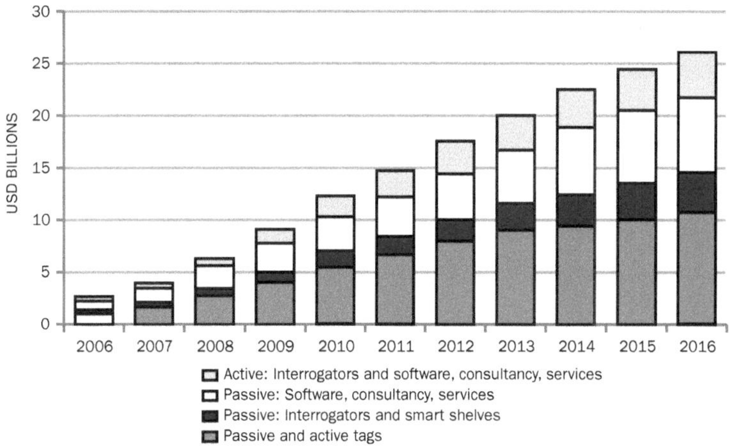

Source: Das and Harrop (2006).

Information technology (IT) is one of the most important resources in creating organizational value through its capability to transform the nature of products, processes, organizations, industries, and even competition itself (Porter and Millar, 1985). Owing to its 'MOST' (mobility, organizational, systems, and technologies) characteristics, RFID has received considerable attention and is considered to be the next wave of the IT revolution. An RFID can allow any tagged entity to be mobile, intelligent, and communicate with an organization's overall information infrastructure (Curtin et al., 2007)

RFID is one of the fastest growing and most beneficial technologies being adopted by businesses today. Adoption of this automatic data collection (ADC) technology has recently been fuelled by the establishment of key standards, retailer, and government mandates, improved technology performance and falling implementation costs. RFID offers great value for many industries and applications (Wyld, 2006). RFID has evolved into a reliable, cost-effective technology used for personal identification, asset management, security, shipping and receiving, inventory control, and many other operations. Improved performance, falling prices, and developing standards continue to move RFID into the mainstream and have made it practical for many organizations to consider its use. A wide range of potential applications is emerging, which could, for example, revolutionize the supply chain, change the way consumers shop, and improve security and safety practices in many industries. But a variety of parties, including businesses, standard bodies and legislators, need to cooperate to ensure that RFID fulfills its potential.

The barcode, a previous identification technology advance, has had a significant impact during the past 30 years. Now, with the emergence of RFID technology, new opportunities will arise and its impact will continue to grow as new applications are realized. In addition, we expect the time to reach significant impact will be half of what it was for barcode technology. Information is the fuel that drives the economy and the society today. The information fuel we use is about to get much richer and more potent with the help of RFID technology.

2

RFID for libraries

Introduction

Globally emerging knowledge-based societies of the twenty-first century will need information to sustain their growth and prosperity. With intellectual capital as investments, knowledge and information have become wealth generators. In this scenario, who can deny the importance of libraries, which are repositories of knowledge resources? A library stacked with books and other information dissemination processes has a physical presence. A library is an institution of knowledge acquisition and learning; it provides invaluable service to its members, patrons, and to a wider community (Ching and Tai, 2009).

Libraries are interested in saving time, money, and labor. Libraries, however, often place excellent public services as their first priority, while still looking to save money and time. Shrinking budgets may dictate that fewer resources will be available for providing the same amount of services. Libraries are searching for ways to control costs, but without having to compromise their lifeblood – providing free materials and a full range of superior services to their users. Currently, technology is most often the hoped-for solution to tightened budgets. When budgets fail to keep pace with use or inflation, less staff are used to do more. Libraries hope that increased use of various technologies can help them to continue to provide the same level of service without increasing staff. Implementation of technology, upgrades, and breakthroughs promise better performance, more efficiency, faster service, and less cost. The emergence of global community as a reality and the accompanying technological developments brought major changes to the library and information sector. User-oriented service is very important for libraries and it is well described in *The Five Laws of Library Science* advocated by the famous Indian library scientist S.R. Ranganathan (1957). They are: (1) books are for use, (2) every reader his book, (3) every book his reader, (4) save the time of the reader, and (5) the library is a growing organism. Even though these laws consider only the books and thus look old, they are still new and applicable even now as we rephrase them by replacing 'book' with 'information,' 'material,' or 'service.' In order to be user oriented, libraries have been growing, or changing, continuously by introducing up-to-date technologies. Due to the advancement of network society, such library services are requested to be extended so that they are shifted to be more network based, more collaborative with other libraries, hopefully providing 24 hours, and so on. In order to realize such services, libraries and library systems should change themselves to be more network oriented and automated ones so that users are able to access their libraries at any time from wherever they want (Minami, 2009).

Libraries have a history of embracing new technologies to improve access to collections and services for customers and internal operations for staff. From the early days of microfilm and long-playing recordings (LPs) to the online catalog and reference databases to internet access and digitization of print and images, electronics have impacted every aspect of day-to-day library operations. A review of the last 25 years shows clearly that libraries and their staff have been avid consumers of technological advances. The modern library is very dependent upon computers for providing access to information sources, including the catalog or catalogs of the library's physical holdings and electronic subscriptions. All the activities associated with lending of stock will normally be dependent on computers and the activities associated with administration, including purchasing of materials, will require computers; most staff will use computers in some way or another in their daily work (Guy, 2007).

What emerged from the 1980s onwards was dependent on what had happened before. A key development, which still has a major effect on libraries today, was the creation of the machine readable cataloging (MARC) format. Towards the end of the 1970s, more affordable computers in the form of minicomputers began to become available and with them emerged companies who spotted that there was now an opportunity to develop library management software (LMS) to run on the new machines. Such systems were referred to as turnkey, standalone systems. Throughout the 1980s, minicomputers improved dramatically in performance and the companies, in competing for market share, regularly improved the functionality of the software. Acquiring such a system for library management became essential for all libraries of a reasonable size. The emergence of the Internet was a very significant development in technological advance for libraries. In the 1990s, there was a major step forward with the emergence of the World Wide Web (WWW). In contrast to the Internet, which is usually defined as a network of networks, the WWW refers to a collection of interconnected documents and resources linked by hyperlinks and uniform resource locators (URLs). Again, librarians eagerly embraced the WWW, developing local websites providing information about services on offer and directing users to resources available in remote locations.

Another key development in the period was that of the microcomputer (later termed 'personal computer'). The Apple II emerged in the late 1970s, but microcomputers started to take off in a significant way with the release of the IBM PC in 1981. Word processing software had initially been developed during the 1970s to run specialized programs on mainframe computers; but as personal computers became more widespread, the software was developed to run on them. The result of these developments, as far as libraries were concerned, was that the specialized terminals used with the early library management systems were replaced by microcomputers. Since microcomputers were multifunctional they could be used for backroom activities (word processing, spreadsheets, etc.) as well as for connection to networks. E-mail that today is a ubiquitous mode of communication in libraries, is not, as might be thought, a development of the 1990s. Rather, it can be traced back to 1965 when multiple users of a time-sharing machine were able to communicate with each other.

New technology always seems to be a hot topic in the library world. During the 1950s, libraries moved from a card-based circulation system to a computer-based system (Cadoo, 2004). This improved privacy with borrowing details now being moved from the public arena (in circulation cards stored in pockets within specific items) to the safety of the library's databases (Schuyler, 2004). With the adoption of barcode technology many libraries were able to streamline workflow within circulation services. The barcode allowed a quicker and efficient way to retrieve a unique identifier, which assisted libraries to track items. Current library management systems use barcode technology and security strips. Using barcodes, a library management system can keep records of lending, borrowing, and shelving status of items such as books, audio or video tapes, CDs, DVDs, etc. Security strips on library items tag their movements. But barcodes and security strips have their limitations. They are slow to read and are prone to sabotaging by thieves. All these lead to irreparable loss to a library and its valuable inventory stock. This is where RFID technology can come to the aid of library managers and users (RFID4u).

RFID, a technology that was originally developed for the military during World War II, has now encompassed all walks of life by making its foray into retail, medical, education, automotives, fast food, the travel industry, and so on. New uses of RFID are being discovered every day. Using them in libraries is just one among them. With the introduction of RFID, the management of libraries is improving dramatically. In libraries, RFID plays a crucial role in checking the catalogs and allowing the customers to pass by card readers that automatically updates the computer system by synchronizing the ID information on the patron's library card with the book and indicating that the book has been checked out. Thus, it ensures a higher degree of accuracy in record keeping, which is a very significant aspect of library management. Moreover, it is expected that the new technology will help in reducing the costs and improve the tracking of library material as it moves in and out of the shelves. Some of the latest RFID products designed specifically for libraries possess some powerful features such as providing real-time information on returned and new items, fast sorting and return of reading materials, providing self-checkout of materials, and allowing web-based monitoring (Gupta, 2008).

Need for RFID

There has been a desire for efficient collection, organization, description, and dissemination of knowledge since the first libraries were established. As library collections and the number of patrons grow, library staff often face increasing demands without the benefit of increasing budgets. There is often not enough time to have any kind of quality interaction with patrons. Technology is the key in approaching these ends, and enhancing the relationship of a library with its patron and its staff (Brown-Syed, 1998).

Much can be gained by using technology in library work. Library automation makes material easier to find and searching more convenient. For many years, libraries have used a combination of technologies to reduce the likelihood of theft, improve stocktaking, and speed up issue and return procedures. Some of the factors that influence the decision for using RFID-technology-enabled tools in user services such as circulation and inventory are as follows.

- *Inefficient use of trained staff.* Generally, the workload of librarians is quite heavy. Even worse, most of 'working with customers' staff time is spent in the check-in and checkout process, while only a small proportion of time is spent on high value-added customer-contact programs. It is in fact an example of ineffective use of human resources. The unnecessary workload on low customer-end services reduces the value of libraries.
- *Need to improve customer service.* Traditional libraries fail to utilize well-trained librarians to foster high value-added customer service. Moreover, the roles of libraries are changing in the new millennium, as patrons demand more high-end service in today's knowledge economy. Professional librarians are trained with knowledge and skills more than doing just book check-in and book sorting. Organizing educational activities and involvement in community events that focus more on customer-relationship activities are becoming their priority.

- *Rising labor costs.* No matter what kind of library, academic or public, as a reservoir of knowledge, they have a high level of books and records archive storage. Managing such a large amount of information is not an easy task that demands well-trained labor, which incurs tremendous labor costs.
- *Increased material handling time and cost.* With an increase in the variety of services offered by libraries, the amount and types of materials handled in libraries increase too. In the past, while books, records, and archives were the main categories of libraries' materials, nowadays, various types of digital documents such as CDs and DVDs increase the amount of libraries' inventory with the transformation into the digital era. Since different kinds of materials have different ways to be managed, substantial time and costs are incurred.
- *Long checkout lines.* The limited space in libraries results in a limited number of checkout stations. Long queues and long waiting times annoy patrons very much. It decreases customer satisfaction and discourages their usage of libraries.
- *Mis-shelved books.* It is prevalent that people after using libraries materials intentionally or unintentionally put them back in the wrong places or on the wrong shelves. In addition, human error is another cause of the problem due to the heavy workload of librarians and the colossal size of libraries and the amount of books that need to be manipulated. In the worst case, the mis-shelved books can only be discovered and replaced correctly in the annual book inventory check. To patrons, it is very frustrating that they cannot find their requested books at the place shown in the

system. Consequently, information in the library cannot be used efficiently.

- *Unreliable theft-detecting system.* Current library systems mainly make use of barcodes for identification. Yet, barcodes cannot be the signal feed for anti-theft systems. The usual combination is to have a combination of barcodes for identification and some electromagnetic (EM)-based anti-theft system. In self-service stations, the identification system must interact with the anti-theft system so that items have the correct EM state after the circulation transactions, that is, a checked out item must be deactivated and returned items activated. These interactions are still unreliable and lacking in functionality and it is prevalent that keys and metal belongings trigger the system and the false alarm causes unnecessary annoyance.

The driver for today's libraries to adopt RFID is the need to increase efficiency and reduce cost. Automation and self-service can help libraries of all sizes toward achieving these aims, and RFID has the added advantage that it can also provide security for the range of different reading materials in libraries. The technology can also improve circulation and inventory control, which helps to optimize the allocation of labor and financial resources. This means that libraries can relieve their professional staff of routine work and operational tasks. RFID can greatly reduce the amount of time required to perform basic circulation transactions. A fully tagged collection can allow the user to use the self-check option, and if available, the self-return option that may contain an optional sorting unit. Because the RFID tags can be read through the item, and also utilises an 'anti-collision' feature, many items may simultaneously be checked out. The anti-collusion feature prevents the information stored on the tag of one item from being mistaken with that of another. In addition to checking out the items, if the security component is active, the security bit will be turned off and the item is recognised as checked out and will not set off an alarm when leaving the library building. This transaction may occur at a self-check station or at the circulation desk. If performed at the circulation desk, it reduces the number of repetitive steps involved in the transaction, such as scanning the barcode and desensitizing each book. As this circulation process requires less physical handling of the material, there may be positive implications for use with archival or special materials. Some systems are equipped with a self-return unit as well, and this unit may have an optional sorter. If purchased, and if the library location information is stored on the tag, the sorter will enable the users to return items at any time to the designated area(s), and the system will do the initial sorting of returned items. Libraries may also benefit from RFID technology through enhanced inventory control. With the aid of a portable handheld reader/scanner, a library staff member may examine a particular stack section and determine if items are either missing or out of place. This task can be accomplished without the need to tip or remove the item from the shelf. With RFID in place, staff may be freed to work on other tasks in the library, including assisting patrons. The technology also has the potential to lower the risk of repetitive motion injuries, as staffs are no longer required to open the item, scan the barcode and desensitize the item. What would have been a three-step process can now be achieved in one step (Dawes, 2005). RFID tags also allow the use of a single label for both circulation and security. When newly acquired media are entered into the collection, only one intelligent tag needs to be attached. This cuts manual effort per operation in half. RFID tags can be used for status control as well as for security checks. There is no need for an additional security stripe.

RFID-enabled library environment

RFID plays a vital role in redefining the library process to make everyone's job easier right from patron to library staff. RFID provides a platform to automate most of the processes performed by the library staff such as check-in/checkout, sorting, stock management, and inventory. In the RFID-enabled library world, the traditional barcode is replaced with an RFID 'tag' and the barcode scanner is swapped for an RFID reader. RFID streamlines operations of the library by:

- enabling self-checkout;
- decreasing time associated with the manual check-in and shelving of returned items;
- quickly locating 'missing items' in the stacks;
- ensuring that new library items reach the shelves quickly.

A typical RFID system for the library environment consists of:

- tags;
- readers and antennas;
- middleware;
- host computer;
- application software (ILS).

Figures 2.1 and 2.2 provide a functional overview of RFID-enabled library environment.

The tag is programmed with distinctive information about the item, which can be directly imported from an integrated library system (ILS) at the tag programming station. It is

Figure 2.1 **RFID-enabled library environment**

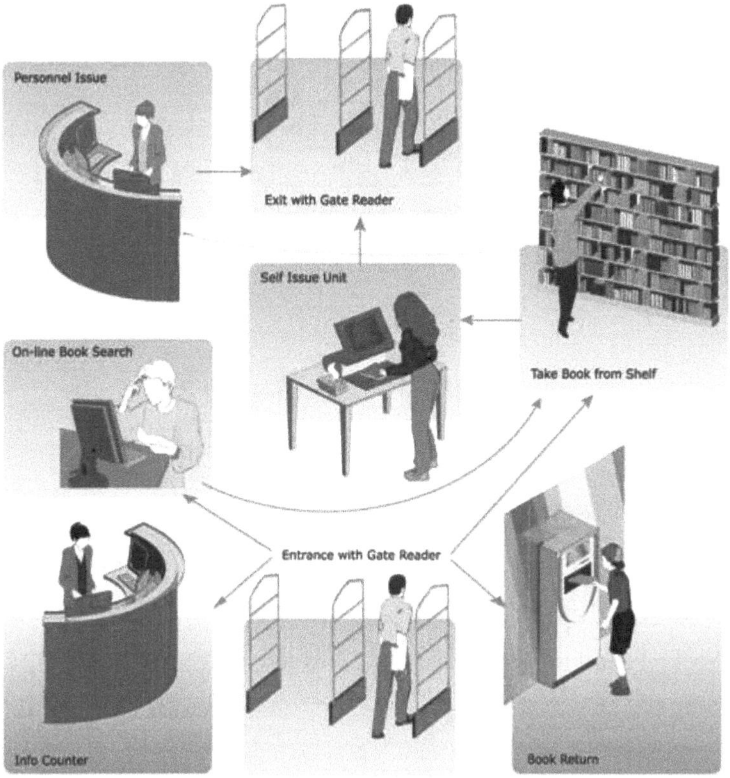

Source: Kern and Nauer (2008).

possible to include many types of information on the tag (such as book title, patron circulation information, date/time stamps), but a tag would typically contain only barcode information. Security can be handled in two ways though sensor gates. Sensor gates can query the ILS to determine an item's security status, or the tag attached to an item may contain a security bit that would be turned on and off by circulation or self-checkout reader stations. Security gates can then detect whether or not the item has been properly

Figure 2.2 **Functional overview of RFID-enabled library environment**

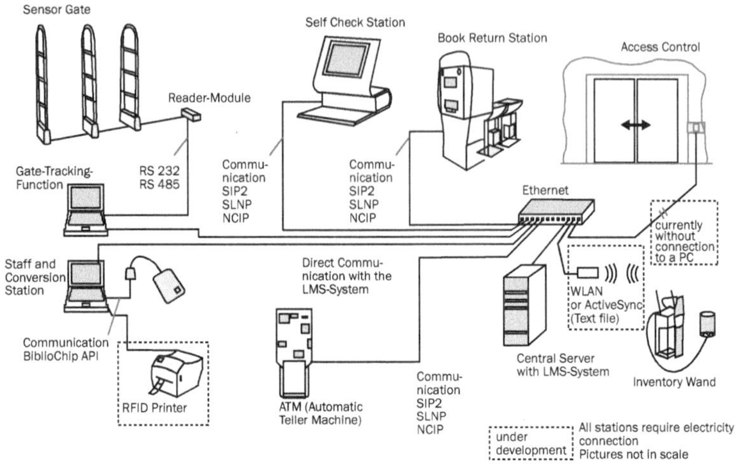

Source: Weber (2008).

checked out of the library. When users return items, the security bit is reset and the item record in the ILS is automatically updated. In some RFID solutions, a return receipt can be generated and materials can be roughly sorted into bins by the return equipment. Inventory wands provide a finer detail of sorting. This tool can be used to put books into shelf-ready order. It can also be taken to the stacks to detect out-of-place items (Smart, 2005).

As general overview of the above RFID components was provided in Chapter 1, we will look at the library-specific RFID components in the following pages.

RFID tags for library

In library systems, the tag is designed as an RFID label that comprises four elements, namely, the chip, the antenna on a foil, the cover paper or plastic label, and the silicon liner (Erwin and Kern, 2003). Passive tags, typically used in library application, gain electric power through an inductive field generated by a reader. Passive tags are smaller, cheaper, and have a longer shelf life compared to active tags. Passive tags do not broadcast the information stored on the tag unless they are requested to do so by a reader.

RFID tags for libraries come in a range of sizes with the most common measuring approximately 50×50 mm^2 and having a paper backing on one side and an adhesive layer on the other side. Special RFID tags are available for CDs, DVDs, and video cassettes. The tags can usually be overprinted with the name and logo of the owning institution if required. The RFID tag may not only be used as a barcode replacement but also contain additional information such as the library and branch identifier, material type and even the title if required. In many cases the RFID tag is also used for item security and so takes the place of the electromagnetic security strip as well (Butters, 2006). Read write (RW) tags are most commonly used in library applications because they allow the library to record changeable information on the tag. Most of the library applications today use high-frequency tags (13.56 MHz) with a read range of approximately one meter.

RFID reader for library

There are many different types of tag readers or sensors available for library applications. Typically, the reader is responsible for generating the electrical impulse that causes the tag to be read because the tags used in libraries are usually 'passive.' The reader interrogates the tag, which then replies with the information stored on the tag. Some readers store the information captured from the tags, while others capture the information and immediately pass it through to the database or ILS. In a typical library, readers are configured to identify tags for purposes of circulation, which include inventory management and theft control. Readers in the library are either fixed or handheld and are located in the library as listed below (Boss, 2007):

- Tagging station is used to program the tags to be attached to the library materials.
- Staff workstation at circulation is used to charge and discharge library materials.
- Self-check station is used to check-in and checkout library materials without staff assistance.
- Book-drop station is used to automatically discharge library materials and reactivate security.
- Sorter and conveyor automated system for returning material to proper area of library.
- Sensor gates are used to verify that all material leaving the library has been checked out.
- Inventory manager is used for inventorying and verifying that material is shelved correctly.

Figure 2.3 depicts the various functional components in the library where RFID readers can be used.

RFID tagging station

In order to convert to an RFID system, all library material must be tagged and programmed. Tagging stations, also called programming stations, are used to affix the tags

Figure 2.3 RFID-based library system

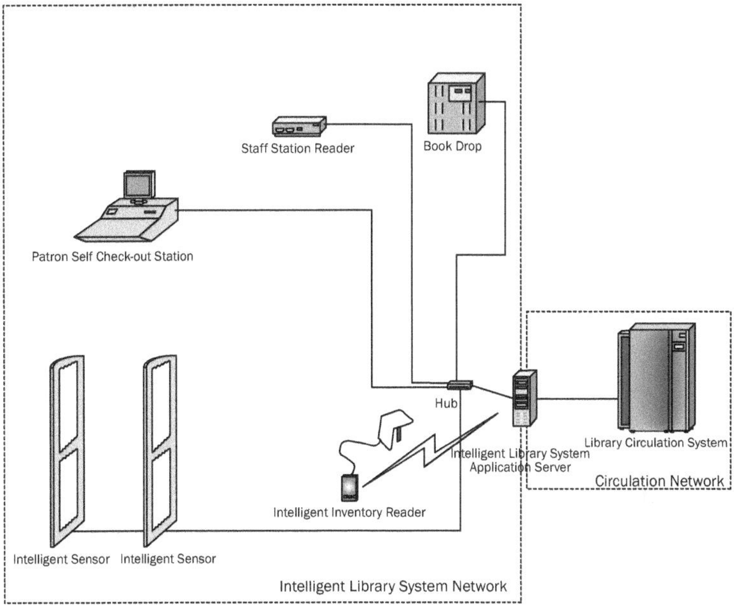

Source: Checkpoint Systems.

to the items and program them (usually via the existing barcode). Programming involves assigning unique identification numbers to library materials with other information (optional) for library transactions (checkout, check-in, and inventory) and activating the theft bit for security. Tagging stations consist of a network PC, reader (Wi-Fi/Ethernet/USB), and antenna.

RFID staff workstation

Serves as a multiple function station – tagging station, circulation station, etc. Staff workstations can be used for checking out, checking in of library materials by library staff and can also be used as a tagging of materials. In case of problems with tagged items not being read at the self-checkout station, etc., instead of engaging the self-checkout station to identify the problems that will delay the checkout process by other users, the staff workstation can be used to monitor and to check the status of the tags (theft bit status, whether alive, etc.). Staff workstations consist of a network PC, reader (Wi-Fi/Ethernet/USB), and antenna.

RFID patron self-checkout station

Allows patrons to self-issue items without assistance from the library staff. Complete with a touch screen and receipt printer, the patron self-checkout station helps libraries to manage the human resources efficiently. Readers placed at the circulation desk and at self-check stations allow staff and patrons to pass stacks of books over the reader, which detects all the tags inside the books. It is not necessary for each item to be scanned individually. Self-checkout stations consist of a network PC with touch screen monitor, RFID reader (Wi-Fi/Ethernet/USB), and antenna suitably placed in an enclosure as shown in Figure 2.4 opposite.

RFID book-drop station

It provides libraries with the best service possible to its members. The library can place the drop box (self-return drop box) anywhere in the campus, for example, canteen, entry gate, etc. and members can drop the books/resources in them and take the return receipt for the same, without visiting the library. In the return process online, the library will get online updates into its resource database. Libraries can make arrangements for collecting the books at any given point of time during the day. Various models of book-drop stations such as wall mounted or stand-alone are available. Typical

Figure 2.4 **Patron self-checkout station**

Source: www.vtls.com

book-drop stations consist of a reader (Wi-Fi/Ethernet/USB) connected to the ILS and antenna with optional sorting units. A PC with touch screen may also be connected so that users can interact. A sample stand-alone book-drop station is given in Figure 2.5.

Figure 2.5 Book-drop station (stand-alone)

Source: www.gaylord.com

RFID sensor gates

Readers built into the sensor gates check the tag to ensure that the item was checked out. If it was not checked out, the alarm sounds (Figure 2.6). This process can be handled in

Figure 2.6 RFID sensor gates

Source: www.gaylord.com

one of two ways: some vendors use a 'theft bit' to indicate whether an item has discharged, while others require that the ILS system be queried as the patron leaves the library (Dorman, 2003).

Sensor gates are used to detect RFID tags that are equipped with electronic article surveillance (EAS). It can detect the

Figure 2.7 RFID sensor gates with surveillance station

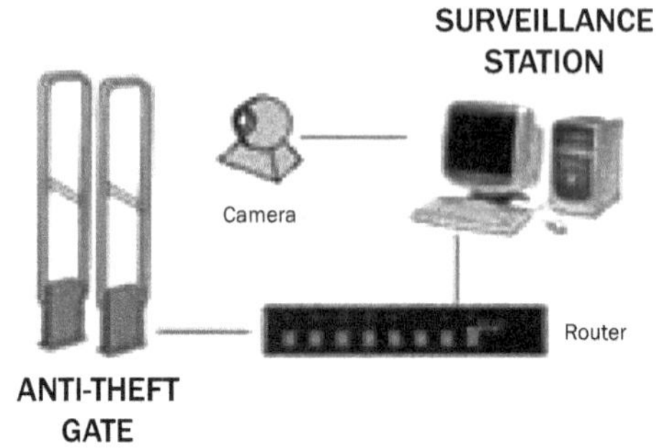

Source: www.vtls.com

RFID tags within one meter range without interference of magnetic items. Upon detecting of armed RFID tags, the alarm will sound with flashing lights on the gate. It has the option to trigger a camera to record patrons that set off the alarm to the surveillance station. The gate performance is designed for fast detection of RFID tags through its multi-protocol reader. The sensor gate itself is a stand-alone unit, which is able to operate independently from the library management system and even the surveillance station. The gate comes with a patron and alarm counter to register the number of patrons passing through the gate and alarm triggers, respectively. However, for better management and monitoring of the data, the gate can be connected to the surveillance station together with the camera alarm triggered recording as shown in Figure 2.7.

RFID inventory manager

A unique advantage of RFID systems is their ability to scan books on the shelves without tipping them out or removing them. A handheld inventory reader can be moved rapidly across a shelf of books to read all the unique identification information. Inventory wands act as readers and as portable databases that store shelf lists. Library staff can use the wands to scan all the items on a shelf and detect which items are out of order or missing. The personal digital assistant (PDA)-based handheld reader complies with multi-protocol and is designed for item level management. This device is ideal to track and monitor assets and for stock management inside and outside the library. The handheld reader is featured with Wi-Fi that enables quick verification and associated action with the help of the enterprise management system. It also works well when connected to a desktop or a laptop via USB. Rechargeable lithium ion and Ni-MH batteries power the handheld reader. Designed with integrated microphone speaker, status LEDs, and single button operations.

RFID middleware

Middleware refers to the software layer that resides between the physical layer components (hardware), firmware, or operating systems, which deal with low-level system calls and communication protocols, and the upper layer stand-alone or even distributed enterprise applications generally interacting via the network. The boundaries between layers are not very sharp, and as software evolves the features of middleware become part of operating systems, firmware, application frameworks, and other layers of the IT infrastructure (Barbosa et al., 2002).

RFID middleware is a new breed of specialized software that sits between the RFID hardware (readers) and the applications. The main goal of this middleware is to process data from tags collected by the readers deployed in the RFID infrastructure or to write ID numbers and/or process data to the tags while commissioning of these tags for assigning to individual items. In addition, it deals with a number of important issues related to avoidance of data duplication, mitigating errors, and proper presentation of data (Prabhu et al., 2006). An application when interfaced directly with the RFID reader has to process a massive amount of data. Such applications also have to take care of different connection-interfaces used by the reader. Handling these different interfaces becomes a huge task, as interfaces require different APIs to program. In order to make applications independent of various types of readers and their connection interfaces, there is a need of an intermediate layer between the RFID reader and the application. This requirement is fulfilled by the RFID middleware. RFID middleware applies filtering, formatting, or logic to tag data captured by a reader so the data can be processed by a software application. This middleware performs the following major tasks (Lieshout et al., 2007):

- It is able to connect RFID readers from different manufacturers.
- Filter the RFID event stream. Holding a RFID tag within the field of a RFID reader produces a flow of events telling that the tag is still in the field. In most cases only the entrance of the tag is important. Therefore all the 'double reads' of the RFID tag must be filtered out.
- The RFID tag contains in most cases only a (U)ID, a unique number. The existing applications know nothing about these numbers, so they must first be converted into known references of the object the tag is attached to. For instance in the case of a library book, the accession number.

Middleware handles the coordination operations, which are a part of the message-oriented middleware (MOM) among readers and automation systems. Besides using messages to pass requests/responses, the middleware also includes security, data broadcast, filtration, route, and error correction to ensure information accuracy and transparency. No matter how library automation systems are enhanced or replaced by other systems, increase, or exchange readers, systems also operate continually without any modification. Depending on the scope of application, RFID middleware includes the categories described below (Yu, 2007):

- *Application middleware.* The development of application middleware is to integrate or connect with readers in the initial stages. The majority of this middleware is provided by reader manufacturers and only offers simple API functions to communicate with systems and readers.
- *Infrastructure middleware.* These kinds of middleware not only include basic functions of data collection and filtration, but also possess the connection requirements of devices to applications, and even provide platform management and manipulation.
- *Solution middleware.* Solution middleware provides the communication interface among automation systems, readers, and RFID.

RFID-based library application software

The RFID-based LMS facilitates the fast issuing, reissuing, and returning of books with the help of RFID-enabled modules. It directly provides the book information and library member information to the library management system and does not need the manual typing. It also provides monitoring and searching systems. The monitoring module will continuously monitor the movement of books across the gates, so that the books taken out without proper checkout will be traced easily and will alarm the librarians. The searching module provides for the fast searching of books using the RFID handheld reader. The physical location of the books can easily be located using this module.

RFID-based library host system or server

The server is the communication gateway among the various components. It receives the information from one or more readers and exchanges information with the circulation database. Its software includes the SIP/SIP2 (session initiation protocol) or national circulation interchange protocol (NCIP) and applications programming interface (APIs) necessary to interface it with the integrated library software. The server typically includes a transaction database so that reports can be produced.

RFID advantages

RFID's property of non line-of-sight operation can be very useful within a library. If we replace the barcode traditionally used to identify library material with an RFID tag, we are immediately able to process multiple items simultaneously and the tag can be located inside the cover in the case of printed material. No longer does every item require individual handling. While users previously had to see library staff in order to checkout any library books, all it takes now is a visit to the self-checkout station to get the books checked out by themselves at any time. The benefits can be realized in improvements to productivity, service, materials handling, collection management, etc. Enhanced security precautions are also possible with RFID as readers at entrances and exits can monitor for unauthorized borrowing of library materials. In each and every activity within a library, RFID technology will provide a greater amount of efficiency and error-free functioning. Libraries thus can obtain several advantages:

- quick check-in/checkout of items for convenience of members and reduce queues at the front desk;
- decrease repeatable tasks;
- increase interaction with patrons;
- extend internal security;
- lower the cost of manipulating and managing collections;
- quick and correct shelving of items that raise the efficiency of shelf arrangement;
- quick inventory check.

The tracking of library assets and their loan processing is another very time-consuming process. Though traditional barcoding systems do help to improve the process to some extent, it is only RFID technology that can offer various enhanced features:

- *Efficient processing.* As each library item contains an embedded RFID tag on the printed label, its availability can be tracked much more efficiently than with manual updating. Library items can be checked in or checked out much faster than the outdated manual processing. In fact, with RFID, processing of returned items no longer requires any human intervention at all. This allows libraries to offer certain facilities around-the-clock, without incurring high costs.
- *Security.* If a tagged library item has not been checked out, any attempt to remove it from the library premises will be detected via the RFID antenna at the entrance gate.
- *Inventory management.* Book inventory processes that previously took weeks or months to execute can now be

> shortened to mere hours using RFID tagging. Wearing a wrist wand RFID device, a librarian only needs to walk through a corridor of bookshelves to check the status of the books available. The RFID reading device reads item information from the books' RFID tags and then automatically interfaces with the library inventory software system to update the databases.

RFID helps improve several aspects of library operation and the experience of all those involved in managing, running, and using the facilities (RFID4u).

Benefits to library management

- Improved security within the library.
- Efficient collection management system.
- Improved collection security.
- Flexible staff schedules.
- Labor saving methods free the staff to help customers.
- Higher customer/patron satisfaction levels.
- Improved inter-library cooperation.
- Better preservation of inventory because of less handling by staff.
- Same security and labeling formats for all items such as books, CDs, and DVDs, hence better management of databases.

Benefits for library staff

- Time saving devices free them to help customers better.
- Labor saving devices free them from doing repetitive, physically stressful tasks.
- Can have flexible working schedules.

Benefits for library patrons

- Faster check-in and checkout of all types of items (books, audio tapes, video tapes, CDs, DVDs, etc.) at the same locations.
- More staff available for assistance.
- Quicker service such as payment of fees, fines, etc.
- Better inter-library facilities, more efficient reservation facilities, etc.
- Faster and accurate re-shelving means patrons can find items where they should be, hence quicker and more satisfying service.
- Height adjustable self-check-in/self-checkout tables are liked by children and physically disabled persons who use the library.

RFID disadvantages

As with any technology, RFID has disadvantages too. The potential disadvantages that were raised included possible threats to privacy, the cost and uncertainty of the return on investment (ROI), the uncertainty of the quality of the performance, and potential effects on staff and patron health. Boss (2007) has provided a detailed account on the RFID disadvantages for libraries:

- The major drawback of RFID technology is its cost. This will discourage librarians from implementing a RFID system in their libraries. *But librarians should be clever enough to find a funding agency to face the challenge. Otherwise they would not be able to be on a par with new technologies.*

- It is possible to compromise an RFID system by wrapping the protected material in two to three layers of ordinary household foil to block the radio signal.
- It is also possible to compromise an RFID system by placing two items against one another so that one tag overlays another. That may cancel out the signals. *This requires knowledge of the technology and careful alignment.*
- RFID tags are typically affixed to the inside back cover and are exposed for removal. This means that there would be problems when users become more familiar with the role of the tags.
- While the short-range readers used for checkout/check-in and inventorying appear to read the tags accurately most of the time, the performance of the exit gate sensors is not so good.
- Privacy concerns associated with item-level tagging is another significant barrier to library use of RFID. The problem with today's library RFID system is that the tags contain static information that can be relatively easily read by unauthorized tag readers. This allows for privacy issues described as 'tracking' and 'hot listing.' Tracking refers to the ability to track the movements of a book (or person carrying the book) by correlating multiple observations of the RFID tag. Hot listing refers to the process of building a database of books and their associated tag numbers and then using an unauthorized reader to determine who is checking out items in the hot list.
- The signal from one reader can interfere with the signal from another in the same coverage and affect the read accuracy.

Conclusion

Technology has evolved at an incredible rate in the past decade. It changes the way librarians serve their patrons and perform their work. Appropriate deployment of technology can help improve service quality and provide efficient operations. However, technology is itself neutral. It is just a means to an end. It carries value to libraries only when it becomes a genuine solution for service provision. Therefore, the choice of technology is an important decision because it is the selection of a service transformation tool that is supposed to bring about benefits and improvements (Ching and Tai, 2009). RFID is providing immediate and long-term benefits to libraries in process, traceability, and security. RFID has a positive effect on the day-to-day handling of library materials as items move through the circulation stream. RFID can enhance existing barcode systems by providing additional features such as automatic self-checkout and return while adding security against theft. The availability of self-checkout kiosks results in shorter lines at the checkout desk and enhances the patron experience. The time savings that result from quicker check-in of returned items is even more significant. Enhanced security precautions are also possible with RFID as readers at all entrances and exits can monitor for the unauthorized borrowing of library materials. RFID reduces the number of lost books and creates savings in time spent searching for books shelved in incorrect locations. Performing inventory and finding lost/mis-shelved items are greatly facilitated through RFID. In the past, these activities required removing items from the shelves, handling them to verify the item ID (barcode or accession number), and replacing items on the shelves in the correct place. The ability to identify an item without handling it makes these activities vastly more efficient. The tedious and time-consuming tasks are accomplished far more efficiently using RFID and free library staff time. All these service efficiencies of RFID in libraries will enable library staff to spend more quality time with their patrons to provide direct customer service activities such as providing reference service and other patron assistance. RFID means a more organized library resulting in happier librarians and more satisfied customers. Even though RFID tags may cost four to five times as much as a barcode and magnetic strip combined, the increase in efficiency and functionality provided by the technology is persuasive enough for an increasing number of libraries to implement or consider this technology.

In the last few years, libraries have begun adopting RFID technology to take advantage of these benefits. Just as every item in the past received its own barcode, each item now receives a unique RFID tag. This simplifies several library functions. For one, circulation workers no longer have to either open or reposition an item in order to scan its barcode for check-in and checkout. Instead, the users need to only bring the item within range of the reader and the item is automatically scanned, regardless of its physical orientation. Additionally, an RFID device can read multiple items at the same time, making it possible to simultaneously check out several items in one motion. Taking this idea one step further, RFID can actually eliminate the need for circulation workers to be involved in the checkout process at all. Many libraries that implement RFID technology install self-checkout terminals to allow patrons the ability to checkout library items themselves without the involvements of library staff. Another area significantly impacted by the implementation of RFID technology is collection management. Traditionally, shelf reading requires the library staff to look at the call numbers of each and every item in the stacks in order to verify that everything is shelved in the correct order.

RFID greatly reduces this burden. In most cases, a staff member can quickly scan an entire shelf using a handheld device that instantly sounds an alert when it encounters any mis-shelved items. The device can also be programmed to search for items that are currently listed as missing or checked out. When collecting use statistics, library staff can also quickly inventory items used by patrons before re-shelving them. All of these benefits create another much bigger benefit for libraries: labor savings. At the very least it frees up library staff who usually perform circulation and collection management functions to spend considerable amounts of time on other duties (Boone and Studwell, 2006). The main benefit brought by RFID technology is the increase in automation. Since all those book check-in and checkout tasks can be done automatically, 24 × 7 self-service library is no longer a dream. People no longer are constrained by the operating hours of the library.

3

RFID systems for libraries: a review

Introduction

The evolution of libraries has been influenced by constant changes due to information technology developments. New technologies have always been of interest for libraries, both for the potential of increasing the quality of service and for improving efficiency of operations. One such technology, which is gaining tremendous popularity among the various libraries, is RFID technology since it revolutionizes the way a library operates. Many large libraries around the world have implemented RFID to speed material check-in, checkout, shelf inventory, and security applications. Counter personnel check dozens of books in or out in mere seconds without manually handling and orienting each item. The tags can also be used for theft detection, much like anti-shoplifting technology currently used by retailers. Librarians using portable computers with RFID readers can take inventory and find misfiled materials simply by walking down an aisle of bookshelves. The reader can automatically detect missing materials and alert the operator. With careful planning and implementation, the introduction of RFID has the potential to provide libraries with productivity benefits, new collection management tools, and improved customer service.

However, the most important issue to understand is that everything associated with RFID deployment is changing rapidly. Software, tag and reader designs, standards, and vendors are all in a state of flux, with many vendors and standards groups vying for dominance. The RFID deployment roadmap must be flexible enough to adapt to the changing vendor and equipment landscape. That is not an easy task with so much in flux. The five-step process below takes these challenges into consideration to minimize the impact of variability on the success of RFID deployment (Dempsey, 2004):

- *Establish the business case.* It is important to develop a strong understanding of the technology's challenges and opportunities right from the start. To develop a business case for selected opportunities, a well-defined plan that is clearly tied to library objectives is needed. Defining the system requirements, establishing costs, and developing clear expectations for ROI (e.g. where, how much, how long will it take, etc.) will help accomplish this step. If preliminary research indicates RFID could be a valuable solution for your library, continue to the next phase, planning a realistic RFID roadmap.
- *Build a practical roadmap.* The roadmap for RFID deployment will be very similar to those for other technology deployments, includes sufficient time and any necessary support for unique RFID issues such as tag placement, reader/antennae placement, orientation and attenuation, and environmental testing.
- *Conduct rigorous environmental proof of concept testing.* Before mounting a full-scale RFID implementation, it is critically important to conduct meticulous product tests in a real-world environment. A complicating factor is the changing vendor and technology landscape. It must also be ensured that technology selections will still be appropriate

at the time of deployment. Contingency plans to deal with potential changes will need to be planned.

- *Carefully pilot the proposed solution(s).* To begin the pilot phase, closely integrate the RFID system to the existing software applications in order to truly approximate a real-world environment. Then connect to host systems for actual data transfer. Because actual operations cannot be shut down to conduct the pilot, be sure the two can run in parallel. This may require modifications to existing systems. To the extent possible, pilot tests must be designed to reflect the business case scenarios defined during requirements definition.
- *Roll-out solutions based on ROI considerations.* In order to obtain management approval for RFID implementations, use pilot test results to demonstrate the technology's value to the organization. Be prepared to specify where RFID will add value, how it will benefit operations, and exactly how much savings in terms of operational efficiencies can be achieved from each application. Be prepared to discuss how this value compares to associated costs and risks, and demonstrate in detail how RFID supports organizational objectives. Analyze results from each implementation and adjust roll-out plans for future implementations accordingly. Even with the best business case analysis and thorough roadmaps, there is one item often overlooked that can determine success or failure of deployment – change management. People are naturally reluctant to change. This is usually caused by fear of the unknown that change represents. Resistance can usually be overcome with proper training. Since much of projected ROI may be based on streamlined processes, staff must be trained how to perform the new processes and operate new equipment. Training should explain why the new technology is beneficial and address any fears.

Experience shows that most successful deployment of new technologies are possible when there are clear business cases, a well-defined roadmap and provisions to continually measure against original objectives and make adjustments accordingly. RFID deployments are no different; they just have a few extra technology considerations to address.

Reviews of RFID systems for libraries

Business partners are an intricate part of the RFID implementation. An emerging trend within the RFID industry is that several companies partner together to provide a complete solution. One company provides the software, another provides the hardware, and third one handles the integration. This may be a possible option if one partner cannot fulfill all the requirements. With the growing interest of RFID into the item management arena and the opportunity for RFID to work alongside barcode, it becomes difficult to count the number of companies who enter the marketplace. Many have come and gone, many are still here, many have merged, and there are many new players. There are many different RFID vendors with different areas of expertise. The manufacturers of RFID systems can be divided into manufacturers of complete systems, manufacturers of subsystems, and system integrators. The companies that focus on developing RFID applications tailored for library applications include:

- 3M (USA);
- Bibliotheca RFID library systems (Switzerland);
- Checkpoint Systems (USA);
- Intellident (Great Britain);
- Libramation (USA);
- Nedap (Netherlands);

- ST LogiTrack (Singapore);
- TAGSYS (France).

The companies that will be discussed here are: TAGSYS, 3M, Checkpoint, Bibliotheca, and ST LogiTrack. An overview of their product offerings are given in the following section.

TAGSYS (www.tagsysrfid.com)

TAGSYS was founded in 1996 as Gemplus Tag and incorporated in 2001 with headquarters in Cambridge, Massachusetts, and offices and R&D centers in France, Pennsylvania, and Hong Kong. TAGSYS is an RFID systems company that makes all the necessary components including tags, antennas, and reading stations. Major markets include security and access control, transportation, and asset tracking. TAGSYS' RFID solutions are currently being deployed in a range of highly specialized vertical markets that include textile services, pharmaceuticals and health care, and libraries. TAGSYS has been at the forefront of RFID deployments by supporting some of the first and largest RFID library deployments in the world: the first ever RFID library installation (Singapore's NLB), Seattle Public Library in the US, Shenzhen Public Library in China, Hamburg Public Library in Germany, and the Geneva Public Library. These installations have encompassed RFID-enabling millions of tagged books, periodicals, CDs, and audiovisual material. The following are the library specific RFID products from TAGSYS.

TAGSYS tags

TAGSYS tags under the brand name Folio are dry inlay and fully converted label type and come in the following three types:

- Folio 370-F3 (square) for books and magazines with size 47 × 47 mm^2;
- Folio 370-LI/370SY-LI for books and magazines with size 48 × 80 mm^2;
- Folio 370-CD for CDs/DVDs with size 36 mm diameter.

The tags support ISO 15693, ISO 18000-3-1 protocols with additional features such as EAS and AFI identifiers (Figure 3.1).

Figure 3.1 TAGSYS RFID tags (© Tagsys)

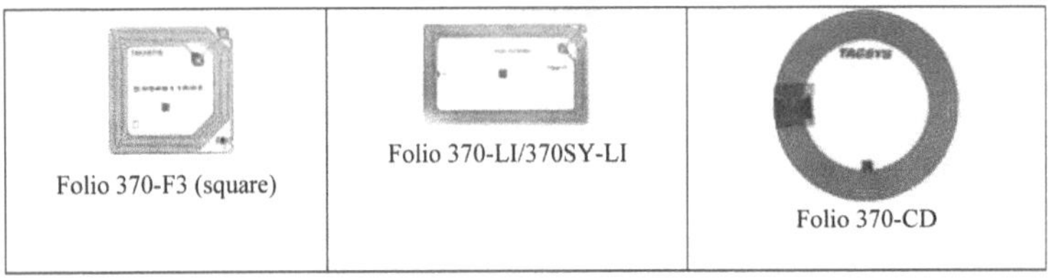

TAGSYS readers

TAGSYS offers RFID readers under the brand name Medio with support of USB, ethernet, and Wi-Fi connectivity. Following are the different readers of TAGSYS (Figure 3.2):

- Medio™ L-P101;
- Medio™ L100/L200;

Figure 3.2 TAGSYS RFID readers (© Tagsys)

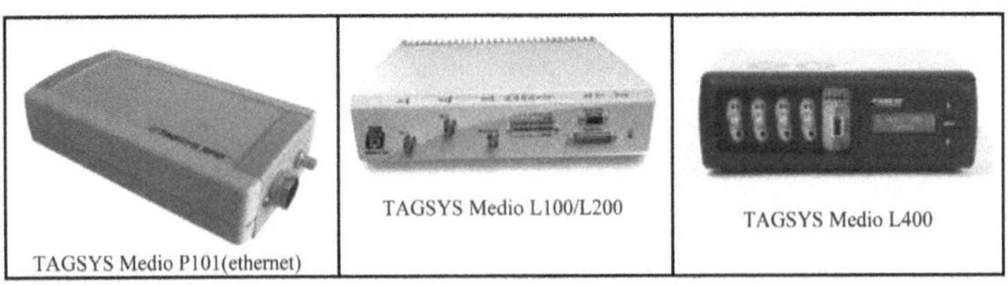

- Medio™ L400;
- Medio™ P013;
- Medio™ P032;
- Medio™ S002/S003;
- Medio™ S004.

TAGSYS antennas

TAGSYS offers essentially two types of antennas – wand antenna and stack antenna. Following are the TAGSYS antennas (Figure 3.3):

- LSA-3 stack antenna;
- LSA-4 stack antenna;
- L-W1 wand antenna.

Figure 3.3 TAGSYS RFID antennas (© Tagsys)

TAGSYS inventory manager

TAGSYS inventory manager consists of L-W1 antenna and one of the medio readers with portable data assistant along with battery case.

TAGSYS EAS gates

TAGSYS offers EAS gates under the name of L-SP that has two versions (L-SP2+ and L-SP3) as of now (Figure 3.4).

Figure 3.4 TAGSYS EAS gate (© Tagsys)

3M (www.3m.com)

3M began testing a prototype RFID library system in 1994. 3M started with a hybrid approach that combined Tattle-tape (for security) with RFID tags (for inventory). In March 2004, it introduced a one-tag solution. 3M manufactures its own RFID tags, with inlays from Texas Instruments, and all tags are ISO 15693 and 18000-3 compliant.

RFID solutions from 3M are integrated ones with built-in readers and antennas. Following are library-specific RFID products from 3M.

3M Tags

3M calls it one-tag RFID system that has 3M RFID Tag D8 for books and 3M RFID Tag CD8 for CDs and DVDs (Figure 3.5).

3M self-check systems

3M offers two types of self-check systems viz., self-check system for both RFID and barcoded items with sorter unit

Figure 3.5 3M RFID tags

to separate both and self-check system for one-tag RFID system (Figure 3.6).

Figure 3.6 3M RFID self-check station

3M tagging station

3M tagging station is called conversion station that includes touch-sensitive screen, optical barcode scanner, and RFID reader (Figure 3.7).

3M book-drop station

The 3M Intelligent Return and Sorter System is designed for libraries interested in automating their check-in and sorting process. Customers can quickly self-return items with 'real-

Figure 3.7 3M RFID tagging station

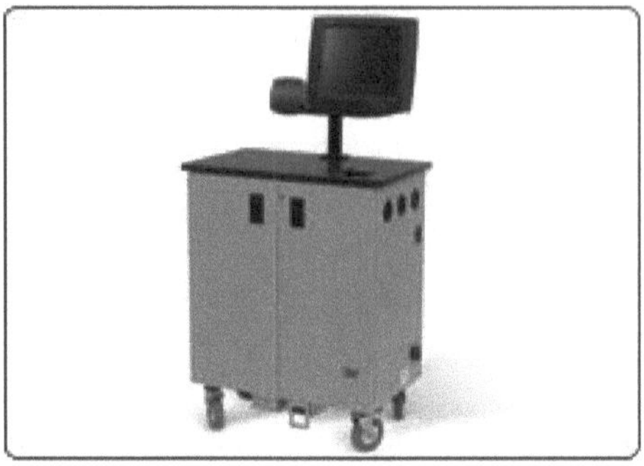

time check-in.' With this feature, they can be assured items are instantly checked in upon return. This allows them to checkout items without exceeding loan limits in the same visit. Designed specifically for libraries, the system can be interior wall or exterior wall mounted, walk-up or drive-up. For libraries with limited space, the system can be used for return only or can accommodate up to three bins with a sort matrix configured to best meet your library's needs. The sweep technology utilized by the sorter for directing items into the bins helps ensure items are sorted properly while reducing damage to items (Figure 3.8).

The 3M Intelligent Return and Sorter System also offers an intuitive user interface that makes it easy for customers and staff to use, including administrative tasks such as

Figure 3.8 3M RFID book return station

configuring receipts, generating statistics, and system diagnostics. The Intelligent Return and Sorter System helps decrease the time it takes for an item to get back to the shelf, helping to increase staff productivity while enhancing customer satisfaction.

3M EAS gates

3M EAS gates are called detection systems and come in different models to support detection of their Tattle-tapes and one-tag systems with options for alarms and people counter (Figure 3.9).

Figure 3.9 3M RFID EAS gates

3M inventory manager

3M inventory manager is called 3M Digital Library Assistant with built-in antenna and reader and data collection equipment that can hold more than one million items during stack verification (Figure 3.10).

Figure 3.10 3M RFID inventory manager

Checkpoint Systems (www.checkpointsystems.com)

Established in the year 1969 with a comprehensive portfolio of products and services ranging from traditional security technologies to real-world RFID systems, source tagging, and service bureau capabilities. Checkpoint introduced RFID solutions in 1998. Checkpoint Systems now sells and services its library solutions through the 3M company.

Bibliotheca (www.bibliotheca-rfid.com)

Bibliotheca is Europe's number one RFID supplier, specializing in RFID solutions for libraries since 2002. With global headquarters in Switzerland and operations in America, Australia, Germany, Italy, and Denmark, Bibliotheca has been the second largest global provider of RFID solutions.

Bibliotheca RFID Library Systems is a spin-off of Lucatron Electronics of Switzerland, a manufacturer of retail surveillance equipment. Founded in 2002, Bibliotheca RFID Library Systems is specialized in the development, manufacture, and marketing of software and hardware for automation and media security in libraries. The BiblioChip RFID System complies with the latest ISO standards as a non-proprietary and customer-specific integrated solution. Bibliotheca's modular complete solution delivers state-of-the-art software as well as all hardware components such as self-check stations, return machines, sensor gates, mobile handheld readers, and RFID labels. Bibliotheca won the Swiss Technology Award in 2005 and the Zug Innovation Prize in 2004. Bibliotheca offers library RFID solutions in the name of Bibliography and has the following products range.

Biblio tags

There are two types of labels for the differing media types (Figure 3.11):

- The BiblioChip Book/Video Label is a label for books, magazines, media packages, audiovisual media, etc. in credit card format.
- The BiblioChip CD RingLabel is especially suitable for CDs and DVDs and can be extended with a BiblioChip CD SecureLabel to provide a complete solution.

Figure 3.11 **Biblio RFID tags (© Bibliotheca-rfid.com)**

Biblio self-checkout

Bibliotheca's biblio self-checkout station comes in different versions to suit varying requirements of the library (Figure 3.12):

- Biblio SelfCheck Jupiter;
- Biblio SelfCheck Mercury.

Figure 3.12 Biblio RFID self-checkout station (© Bibliotheca-rfid.com)

- Biblio SelfCheck Neptune;
- Biblio SelfCheck Mars;
- Biblio SelfCheck Hybrid;
- Biblio SelfCheck Orion.

Biblio tagging station

Biblio StaffStation allows staff to initialize library media and ID cards at their workstation. The large, powerful table-integrated aerial is concealed below the desk (Figure 3.13).

The Biblio StaffStation is fully integrated into the interfaces of the most important library systems, enabling users to work with RFID under their accustomed on-screen interfaces.

Figure 3.13 Biblio RFID tagging station (© Bibliotheca-rfid.com)

Biblio BookDrop station

The BiblioReturn and BookDrop facilities allow media and media packages to be returned swiftly and simply. In keeping with the varying requirements of different libraries, Bibliotheca offers two BiblioReturn devices and two Biblio BookDrops for use by patrons. In most instances, return facilities are combined with a sorting device. The BookDrops generally collect the media in a book trolley; only the BookDrop Secure provides the option of a connector module to a sorting system (Figure 3.14).

Figure 3.14 Biblio RFID book-drop station (© Bibliotheca-rfid.com)

Biblio EAS gates

Bibliotheca's EAS gates are called Biblio gates that offer an impressively high identification rate and powerful performance. Successful identification and safeguarding is independent of media type and alignment of the RFID chip while passing through the gates. Connection of numerous LMS/ILS systems is a key efficiency standard. A specially developed gate-tracker program, which is independent of LMS/ILS automatically, provides information in a log file at defined intervals, as to which media that have triggered an alarm. Several gates can be monitored simultaneously (Figure 3.15).

Figure 3.15 Biblio EAS gates (© Bibliotheca-rfid.com)

Biblio inventory manager

The portable devices allow mobile maintenance of library stocks directly at the shelf. The specially designed, light manual aerials read all media chip data based on a predefined reading range. A handy PDA facilitates on-site inventory control. Misplaced and reserved media can be tracked down according to various search criteria using acoustic and visual display (Figure 3.16).

Figure 3.16 Biblio inventory manager (© Bibliotheca-rfid.com)

ST LogiTrack (www.stlogitrack.com)

ST LogiTrack Pte Ltd, established in January 1998, is a joint venture company between Singapore Technologies Logistics

Pte Ltd and Singapore Technologies Electronics Ltd. ST LogiTrack offers the library RFID solution under the brand name EliMS (Electronic Library Management System). EliMS consists of the following RFID-based library system components.

EliMS borrowing station

The ELiMS borrowing station is a self-service station. Simple instructions are presented to the patrons to enable them to checkout items at their own convenience. Instructions can be configured in different formats and different languages to suit the specific requirement (Figure 3.17).

Figure 3.17 EliMS borrowing station

ELiMS book drop

The ELiMS book drop allows patrons to return their items at convenient locations and to have their loan records updated instantaneously (Figure 3.18).

Figure 3.18 EliMS book-drop station

EliMS Remote Return Kiosk

The EliMS Remote Return Kiosk operates independently and can be located anywhere, inside a building, in a public transport terminal, or even along the road; much to the convenience of patrons (Figure 3.19).

Figure 3.19 EliMS Remote Return Kiosk

ELiMS sorting station

The ELiMS sorting station allows items to be sorted and returned quickly to the shelves so as to reduce turnaround time.

ELiMS EAS gates

The ELiMS EAS gates are the anti-theft part of the integrated library management solution using the same RFID tags embedded in the library item (Figure 3.20).

Figure 3.20 EliMS EAS gates

ELiMS counter station

The ELiMS counter station is a backup station that performs the borrowing, returning, and disarming functions. The librarian can use the system in the event when some of the stations (i.e. borrowing, returning, etc.) malfunction or when other activities such as payment of fines for late return of items are necessary.

ELiMS administration station

The ELiMS administration station monitors the status of ELiMS components, such as borrowing station, sorting station, etc. This station has features such as remote reset, transaction logging, status update, etc.

ELiMS multi-purpose station

The ELiMS multi-purpose station is a self-service station. It can be used as a borrowing station, item return, and/or sorting station. It is very suitable for small libraries with low patronage and small collections of library materials.

ELiMS inventory manager

The ELiMS stock management system makes locating and identifying items on the shelves an easy task for librarians. It consists basically of a portable scanner and a base station.

Discussion

Many libraries around the world have tried integration of RFID technology in the business process of library. Some have learnt the hard way and failed, some have succeeded by learning from failures, others learned from the failures and successes of the early adopters and implement the best RFID solutions. Choosing the right RFID system for the right library applications can be a frustrating experience. The best RFID solution for a library is actually the one that seamlessly integrates RFID technology with the existing library management system.

While library RFID systems have a great deal in common with one another, including the use of high-frequency (13.56 MHz), passive, read-write tags, there are some significant differences. Some vendors offer every component to make a complete RFID system, while others are resellers or supply only hardware for specific functions. As seen in the previous sections on products offering from RFID companies, TAGSYS offers individual RFID components such as tags, readers, antennas, and gates, and the system integrators assemble these components as per the requirements of the library applications whereas other companies, 3M, Bibliotheca, and ST LogiTrack offer functional specific modules such as tagging station, self-checkout station, and book-drop station with pre-assembled RFID components. Greater customization is possible with TAGSYS kind of solutions but required trial and error methods to achieve the desired results. On the other hand, pre-assembled solutions that have already been tried and tested may not have enough customization possibilities. For example, there are a lot of options for configuring a self-service unit, and it takes time to determine which settings are most appropriate for a specific library. Table 3.1 provides some of the features that are relevant for product selection and evaluation based on sample quotation received from RFID companies for comparison.

There are many choices and trade-offs for a library in the RFID marketplace, but the key to buying a system that will be a long-term investment is to get RFID tags that meet current standards and can be reprogrammed and used with the majority of RFID readers. This is more like insurance for the investment that the library makes in RFID tags that will outlast its commitment to whatever RFID reader hardware is purchased.

Conclusion

The top priority for most libraries is to find an RFID company that has successfully performed numerous implementations at an organization similar to their own so

Table 3.1 **Comparison of features offered by RFID companies**

Requirement	Check point	TAGSYS	3M	ST LogiTrack
a. Compatible with existing LMS	Through SIP2	Through proprietary NCIP/SIP2	Through SIP2	Using NCIP/SIP2
b. Must work in Linux OS	No	Yes	No	No (.NET)
c. Compliant to ISO 15693 and 18000	Yes	Yes	Yes	Yes
d. No. of items simultaneously checked out/in	5	16	Not provided	3
e. Transmission distance of RFID antenna (gate)	120 cm	91.4 cm	Not provided	100 cm
f. Memory in tags	1 K	1 K	2 K	1 K
g. RFID communication to PC	Serial port	Serial/ USB/ ethernet	Serial port	Serial port
h. Tracking of IDs of library materials that go out without valid checkout	Yes	Yes	Not provided	Not provided

that these organizations can be contacted not only to ask about the performance of the technology but also to enquire how well the RFID vendor worked with the library's ILS vendor, as well as other technology vendors if applicable. Products from different vendors (access control doors) that do not operate together seamlessly require special accommodations that may increase the cost of the overall project. Product compatibility such as the interface between RFID system and LMS is very important for the success of the project. Test the compatibility of RFID with LMS early before making a purchase to avoid costly alterations down the road. Sometimes there is a series of trials and errors required before all settings are just right. The library may want to try a specific setting for sometime to see how it works and then change to find best configuration. Understand the vendor's service terms, guarantees, warranties, and payment options. It is also equally important to institute troubleshooting procedures prior to encountering a technical issue. After equipment has been installed, most technical support will take place remotely. Remote tech support can save time and money, but it can also present communication challenges. With proper training from vendors, a library's technical staff can probably handle minor maintenance and basic troubleshooting, and they become an extension of the vendor's technical support team. Investing a small amount of time to train the library's technical staff in maintenance and troubleshooting goes a long way and help vendor technical support teams identify and diagnose problems quickly and efficiently.

4

RFID implementation in libraries

Introduction

There is never a 'best time' to adopt any new technology – today's facilities are always better, cheaper and faster than yesterday's, but whatever is bought today is almost guaranteed to have been superseded by something even more ideal by the time tomorrow comes. In the end, we just have to proceed when the time seems right for us – when what is available will adequately fulfill the requirements for the short to medium term.

Radio frequency identification (RFID) is currently gathering a lot of interest within many industrial sectors and particularly in the case of a library enabling some of the more simplistic tasks such as checkout and check-in. Libraries have implemented RFID applications in collection management, circulation services, and in inventory operations to employ the functions of identification, rapid response to increase efficiency and accuracy. However, selecting the right RFID technology solution involves myriad options in terms of frequencies, tag configurations, types of antennas and readers, and systems – each with a potentially significant effect on the capabilities of the overall system. For example, systems using higher frequencies to communicate between the tag and the antenna provide more reads per second but can be limited in terms of practical read distance, while lower-frequency systems can read from a distance but can miss many tags. Active tags dramatically increase read distance but cost much more than the passive tags and have shorter lifespans. And antennas that are effective in the lab can miserably fail when deployed in the real world due to unforeseen radio-frequency interference. Choosing and using RFID requires detailed effort, from initial investigation and vendor selection through planning and implementing the conversion to ongoing maintenance and evaluation. While it is common now for libraries to have information technology expertise within their organizations, RFID with its blend of radio technology and electronics may appear unfamiliar and unique. It can be difficult for library professionals to evaluate vendor solutions and to weigh features and benefits since RFID is not one single technology. The way RFID readers communicate with RFID tags varies from application to application, as does the frequency at which they communicate. The good news is that within the library world almost all systems operate at the same frequency and use the same or very similar principles. But the important questions are these (Butters, 2006):

- Do we really know what we want RFID to deliver in our library or can we devise a process to find out?
- Are there systems on the market that can deliver what we want?
- Can we construct a positive and realistic business case to demonstrate the benefits?
- Do we have the skills and experience, or access to them, to make the right system evaluation/selection?

The RFID implementation in general is more than just establishing potential locations of RFID reader and antenna.

It involves requirement analysis, process analysis, technical assessment, budget analysis, technology selection, pilot case, and evaluation. It also involves integrating RFID components into network infrastructure and environmental influence that may affect the implementation. The deployment of RFID requires forming a team, which consists of subject matter experts, applications developers, and communications engineers. The team should work together in all stages. The RFID deployment may consist of the following stages (Al-Muhamed, 2007):

- *Processes analysis.* To establish a baseline by identifying and analyzing the current process called 'As-Is.' The result of the analysis will be used to optimize the processes by eliminating non-added value tasks and redesign the processes called 'To-Be.'
- *Site survey.* To collect information on RF interference, environmental issues, network infrastructure, physical location of the RFID devices and power.
- *Technology selection.* To identify the RFID functionally, frequency selection, RFID tags type (passive, active), data rate, read range, RFID protocol, standards, and middleware selection, centralize management and RFID network security. This stage is considered as the most important and critical stage from the whole lifecycle of the RFID deployment.
- *Implementation.* To integrate middleware to existing applications, install RFID devices, connect devices to networks and perform equipment configurations.
- *Test.* To define success factors that indicate the expected result is achieved by performing different test scenarios starting by testing the readers' connectivity, tag reach by the readers, middleware ability to filter tags, etc.

- *Maintenance.* RFID devices require continuous maintenance and monitoring with updates and new features deployment.

A typical RFID implementation in a library would involve tagging every item in the library (books, tapes, CDs, DVDs, magazines, etc.) with an RFID tag. This tag contains an identification number that uniquely identifies the item. It can also contain additional information such as shelf location, whether or not the item has been checked out, title, author, or any other useful data. Also, circulation stations are equipped with tag readers that can be used to query the database or checkout books. The entrances and exits to the library are equipped with sensor gates that read the RFID tags on the library-owned items when the patron enters or exits the library. An alarm sounds if an item is detected going through the gates without being checked out. The inventory and shelf management are supported with a handheld RFID reader and personal data assistant. A typical RFID implementation in a library is given below as a case study to demonstrate the processes at various stages of implementation.

RFID implementation in IMSc library: a case study

Background

The Institute of Mathematical Sciences (IMSc), Chennai, India is a national institute for fundamental research in frontier disciplines of the mathematical and physical sciences: Theoretical Computer Science, Mathematics, and Theoretical Physics. The major aims of the institute are:

- to create and provide an atmosphere and environment suitable for creative work and pursuit of knowledge and advanced learning in the areas of mathematical sciences;
- to promote and conduct research and original investigations in fundamental sciences;
- to stimulate a zest for creative work and to cultivate a spirit of intellectual collaboration among academic workers in pure and applied branch of the sciences;
- to arrange lectures, meetings, seminars and symposia in pursuance of its academic work and for the diffusion of scientific knowledge;
- to invite scientists actively engaged in creative work to deliver lectures and participate in its academic activities.

The IMSc library is the information backbone of the institute providing access to vast array of information resources that are relevant to the academic and research activities in the areas of Mathematics, Theoretical Physics, and Theoretical Computer Science. In addition to serving the academic and research community of the institute, the library also serves about 5000 walk-in users in a year from colleges, universities, and research institutions from different parts of the country in meeting their academic and research information needs.

In an effort to enhance and maintain its holdings of books, journals, and other sources of information and to provide user-friendly access to these resources, the IMSc library has established the following broad goals:

- to foster and support research, teaching, and learning by providing resources, access to knowledge, information, and ideas;
- to establish and maintain state-of-the-art library facilities;
- to acquire, organize and disseminate need-based information resources in print, non-print, and in electronic forms;

- to design and develop library intranet server to host all the electronic information resources and to provide campus-wide information access;
- to design and develop web-enabled information access tools.

The accomplishments of the IMSc library over the years are substantial in meeting the above goals. Through innovative acquisition, organization, and delivery of information, the library effectively provides information for the users – faculty, students, and visitors.

The IMSc library is committed to:

1. use of technology to enable user access in a global context;
2. designing and refining innovative services to meet user needs;
3. wise and responsible use of fiscal resources through sharing;
4. providing the richest array of information possible;
5. remaining the premier academic and research library in the country with global access.

The IMSc library presently holds a collection of about 65,000 volumes and subscribes to over 300 print journals and provides access to over 3500+ online journals.

The following were scenarios in major functional areas of the library before implementation of RFID system:

- 24 × 7 access to library is enabled using access control system with proximity user cards.
- All the library items (books and bound volumes) were barcoded with accession number as identifier.
- The library uses an integrated library management system on a Linux platform.

- Scanners are used for checkout and check-in of library materials during the staff period.
- During no staff period, library users self-checkout library materials by signing the book card and dropping it into a box kept at the circulation desk and self-check-in by dropping the library materials in a big wooden drop box. Library staff check these transactions next morning and make relevant entries in the library management system.
- Users also self-checkout individual issues of journals.
- Inventory management of library materials is done using the scanner and a laptop computer.

The expected scenarios after implementing an RFID-based system in the IMSc library:

- an environment that enables the librarypatrons to checkout library materials simply bypassing through sensor gates that read the tags and automatically update the library's software system;
- streamlined self-checkout of library materials (books, bound volumes, media, and individual issues of journals);
- streamlined self-check-in of library materials;
- more accurate transaction records;
- the RFID gate to keep track of all the movements of both users and materials;
- tracking of materials that go out of the library without valid checkout;
- streamlined access to the bona fide library access card holders;
- faster and accurate inventory of library materials;
- faster location of misplaced library materials;
- better shelf management.

The following are the various processes followed during RFID implementation in the IMSc library.

Knowledge process

As stated earlier, it is important to understand what RFID can do in libraries and what it cannot. Internet is the best source for getting information about manufacturers of RFID, vendors, and system integrators. It is not enough just to go by what the manufacturers/vendors claim as implementation benefits may vary from different application environments. Case studies that share experiences of RFID implementation in libraries are another important source of information. After gaining a fair amount of knowledge about RFID and the benefits that it brings to the library, cost estimates were to be made as the budget needs to be approved and sourced. Through informal sources like talking to vendors, budget estimates were made. When cost estimates were made for the first time during the year 2004, the passive tags for books were costing little over one US dollar per tag. The tag price made it very difficult to arrive at a workable budget and it was decided that the project may be postponed. After tracking the prices and performance improvements of RFID in libraries for about two years, the project was revisited in the year 2006. By then the cost of tags had come down almost half to 50 cents per tag. Also improvements like network-enabled RFID readers using Ethernet/Wi-Fi in place of USB, READ-ID-enabled sensor gates, etc. were made possible. So, a wait of two years saved about 50 per cent of tags' costs and improved RFID components. On realizing the fact that further reduction in tags' cost might not happen in the near future, it was decided to start the project. It is believed that implementing RFID technology would strategically enhance self-services in the library and at the same time would also provide staffing efficiencies to develop new services such as value added users help and information support services at point-of-need. RFID technology presents the opportunity to put library services at the leading edge of the organization.

Tender process

There are a good number of manufacturers in the world market of RFID systems and each one has unique solutions to offer. In order to arrive at a uniform or near uniform solution to meet our requirements, technical quotations were invited. Technical quotations were evaluated based on requirements and the system capabilities. With this it was possible to arrive at solution/equivalent solutions that most of the RFID manufacturers will be able to meet with. Based on the streamlined technical requirements, commercial quotations were invited. The following are two important points that were kept in mind while inviting quotations:

- support base in India;
- smooth integration of RFID system with existing integrated library system (ILS) or a better cost-effective alternate solution.

A committee was formed to evaluate both technical and commercial quotations. The committee evaluated quotations based on the solutions offered and their relative costs. The RFID hardware was sourced from TAGSYS, France and system integration was done by a LMS vendor based in New Delhi, India.

Implementation process

A project schedule was drawn and the users were informed about the project. Though the project was to complete in six months time, it took almost a year. The implementation involved location of RFID system components, installation, configuration, testing a sample, tagging, fine tuning of hardware and customization of software configuration wherever required, live demo, and putting the system into public use.

System components

The IMSc library purchased 65,000 book tags, 1000 CD/DVD tags, and 500 RFID member cards. The following RFID system components were also purchased:

- one pair of RFID pedestals (sensor gates) with READ-ID feature;
- one RFID tagging station/staff station;
- one RFID self-check-in station/book-drop station;
- one RFID self-checkout/borrowing station;
- one RFID inventory manager.

RFID sensor gates

The important component of a library RFID system is sensor gates. TAGSYS sensor gates are called the L-SP2 EAS system. The components of the L-SP2 EAS system are contained within the L-SP2 pedestal. At least two pedestals are required for each EAS gate. A set of pedestals is known as a gate and may consist of several pedestals. TAGSYS Security Gates do not need to be linked to the library database, and can still operate when the ILS is down or under maintenance. The security gate does not require additional equipment to operate (Figure 4.1). The L-SP2 pedestal includes new functionalities as:

- multi-protocol features, which make it compatible with C220, C320, C370 (Philips SLI) RFID tags;

Figure 4.1 RFID sensor gates at IMSc

- one block system electronic embedded in each pedestal;
- mechanical compatibility with EAS system;
- a warning visual and audible device (LED/Buzzer board);
- one people counter (based on a photoelectric switch couples to a reflector);
- a 'read ID' version including an Ethernet serial interface.

The electronics unit located on the bottom of the pedestal manages the whole system. This electronics unit generates the RF signal transmitted by the antennas and picks up the reply from the TAGSYS RFID tag. If a TAGSYS RFID tag with an active theft bit (ON) is detected, the electronics unit will activate the alarm of the LED/buzzer board on the pedestal.

Pedestals were mounted at 915mm apart (edge to edge) for maximum reliable performance and a clear space of at least 500mm around the edge of the pedestals to ensure that the antennas will not be detuned. The gates are connected to the library network on a pair of IP addresses for easy maintenance, reconfiguration, etc. Detailed instructions on installation of hardware, software, configuration of Ethernet interface for READ-ID sensor gates, synchronization of pedestal communication, antenna tuning for optimal performance, testing and verifying the performance, and troubleshooting are given in the TAGSYS manual L-SP2 user guide (*www.tagsysrfid.com*).

RFID tagging station

The RFID tagging station, also called RFID staff station, consists of Ethernet Medio101 reader, LSA3 antenna, software for configuration, power cable, and a network PC. A scanner is also installed to use the barcode already affixed in library materials for faster and error-free assigning of unique identification to library items (Figure 4.2).

The installation involved connecting MedioP101 Ethernet reader to a computer through Ethernet port and to LSA3 Antenna through the USB port, loading a driver for MedioP101 Ethernet reader, assigning an IP address to the reader, and installation of software for tagging. The operation involves affixing RFID tags on library items, scanning the barcodes of library items to help assigning an error free unique identification number, validating the data, writing the data on the tag and saving it.

Figure 4.2 RFID tagging station/staff station

RFID self-check-in station/book-drop station

RFID self-check-in station also called RFID book-drop station consists of MedioP101 Ethernet reader, LSA3 Antenna, a receiving cart for about 500 books, RFID-enabled lock for secured book-drop slit, a wooden slider, a wooden enclosure, CD containing configuration software and power cables. In addition, an RFID enabled lock for member authorization has also been installed for secured and authentic return transactions. The installation involved assigning an IP address to MedioP101 Ethernet reader and connecting it to the LSA3 antenna. This does not have an interactive unit like self-checkout station. The operation of RFID self-check-in station involves (Figures 4.3 and 4.4):

- user shows identification card to the unit for authentication;
- on member authentication, the book-drop slit opens;
- user drops books one by one (through the RFID enabled slider);

Figure 4.3 Self-service book return/book-drop station

- the system checks in the books and enables the theft bit;
- the books are received in a receiving card kept inside the self-check-in station;
- the library staff removes the books from the cart and reshelves them at their places.

Figure 4.4 Self-service book return/book-drop station (backside view)

RFID self-checkout station

RFID self-checkout, also called RFID borrowing station, consists of Medio101 Ethernet reader, LSA3 Antenna, farm factor PC, a touch screen for interactive sessions, a wooden enclosure, CD containing configuration software and power cables (Figure 4.5). It also has optional printers for printing

Figure 4.5 Patron self-checkout station

the transactions slips. Operation of self-checkout is straightforward, typically:

- borrowers present their RFID enabled identification card to the unit;
- items are placed on the reading table;
- items are read and passed for checking to the library management software (LMS);
- the LMS returns its decision to the self-checkout station;
- any items that may not be loaned will be advised on screen;
- security data is written to the tags to allow or deny them to pass the security gates;
- a receipt may optionally be printed.

RFID inventory manager

RFID inventory manager consists of MedioP101 Wi-Fi reader, inventory wand, personal data assistant (Wi-Fi-enabled iPAQ pocket computer), optional laptop computer, and CD containing configuration software and battery case (Figure 4.6). The same system is also being used for shelf management. The operation of inventory manager for stack verification includes waving the inventory wand through stacks and saving the items scanned into the file, which will then be imported into ILS for verification reports. The operation of the inventory manager for shelf management includes waving the inventory wand through stacks that checks and validates the shelf order, and the system issues a warning if the order is altered with item details so that the staff can put the items in the right order (Figure 4.7).

Figure 4.6 RFID inventory manager with laptop

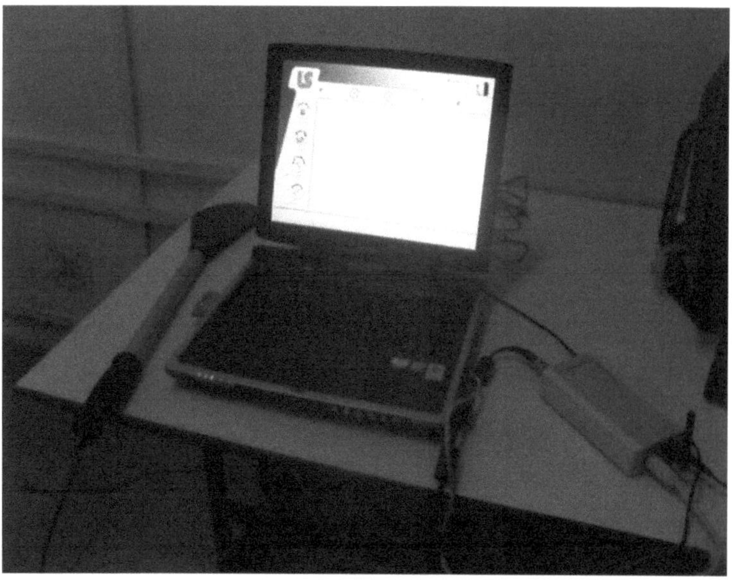

Figure 4.7 RFID inventory manager on the move

Tagging and validating

Tagging is an important task of the whole project. Unless tagging of the target materials is complete, the RFID system cannot be put into use. There were about six people (two in a group – one to check and validate in the system and the other to paste the tag) engaged and tagging of all 65,000 items took only about 45 working days. The pasting of tags is standardized (two inches from the bottom and two inches from the spine) to enable faster inventory and searching of items. On testing, it was found that staggered placements of tags are advisable to avoid tag collision. Each tag has been verified and validated before the item is put back into the bookrack.

Testing and customization

Testing of various RFID components (sensor gates, borrowing station, book-drop station, and inventory wand) were done with live tags. There were several adjustments and customization both at hardware and software levels to bring up the system for optimum performance. These are very time consuming processes and require a lot of support and cooperation of library staff, library users, and the RFID vendor.

Live demo

A demo of the new RFID system has been organized for the users and users were requested to try on their own with the help of library staff to get acquainted with the system so that they can operate it on their own. During this process, several feedbacks were received and incorporated into the system before they were released for use.

Post installation

An RFID-enabled access control system has been installed to facilitate secured access to the library. Every library user has been provided with RFID-enabled member card that also acts as library access card. The access control system is so configured that it will remain disabled during the working hours (staff period) and will be enabled outside office hours (no staff period). This is to facilitate the walk-in users to access the library without having access cards (as there will be hundreds of walk-in users every day) during the normal working hours.

The system meets most of the expectations. There are no circulation desks now and checkout and check-in processes were streamlined to a great extent, which resulted in more accurate circulation transaction records. The inventory of all the library items is done in just a few days' time. A trial inventory check before and immediately after the installation of RFID was done and fortunately there was no loss reported. However, during the full-scale inventory after a year, it was found that there were some books found to be missing. It was later found that these books were missing because of some deliberate attempts by smart users.

Implementation issues

The issues faced during implementation may be specific only to our sites. However, listing them may provide some helpful tips.

- *Tagging*. Library science trainees on a contract basis did tagging of library items. The trainees were not familiar with the existing library practices. There were cases of wrongly pasting CD tags in place of book tags. There were also problems of mismatch of items and records in the database. Placement of tags was initially at a particular position but after realizing the fact that it affects the read rate, placement was staggered.
- *Shelf checkout*. Though the system supports simultaneous multiple items checkout, it was often found that some items were not read. Users find it difficult to identify which

items are not read. As this may defeat the purpose of a quicker and error-free checkout process, the number of items that can be checked out simultaneously was restricted to four. The other frequent problem was read rate and it was found that this is due to the software interface and not that of reader or antenna. The read rate problem was fixed in the new software interface.

- *Book drop*. Since this is a stand-alone book return station with manual sliding unit, the design of book drop was altered at least four times in order to obtain the optimum performance levels. Again here the read rate problems did happen and were fixed in the software interface later.
- *Sensor gates*. The gates' read accuracy has never been 100 per cent and factors causing the read error are more environmental in nature. Tuning of antennas also helped read accuracy. But the serious issue is the correct recording of IDs of items that go out of the gates without proper checkout. New improved electronic units of gates were installed to ensure the READ-ID feature. However, it was found that when READ-ID feature is enabled there is a compromise on the read rate.
- *Inventory management*. Owing to the metal environment of the bookshelves, especially the sides of the shelves and bookends, readers did not identify some tags very often and some not at all.
- *Software interface*. Most of the implementation problems were related to get the software interface between RFID system and LMS customized to meet the local requirements. The interface was implemented initially on a Windows platform and finally a Java-based interface was implemented on a Linux platform. There were several occasions where it was thought that the read rate problem

particularly is due to limitations on the capabilities of RFID hardware components. But it turned out to be problems related to software interfaces.

Lessons learnt

Effective RFID implementation is a challenging task as the technology is still maturing, standards are evolving, ROI remains unclear, and best practices have yet to be formalized. Successful implementation of RFID in libraries involves the following:

- Technical knowledge is required to familiarize with the RFID technology. How does it work? Which are the applications of this technology? Which are the potential benefits it brings to the different systems? Which are the functions that other companies are improving with this technology? Which are the problems or challenges of the technology?
- Cooperation between library and RFID vendors.
- Cooperation of library authority, users, and library staff.

Early RFID implementations were fundamentally driven by external mandates, but along with significant technological improvements, more readily available component options, cost reductions, and shared lessons learned, the technology has proven its value in driving significant operational efficiencies, and RFID has gained a broader adoption. Many organizations encounter similar challenges as they embark on their individual RFID journeys. Leveraging lessons learned can assist in building a technical knowledge base and help appreciate the realistic expectations and physical limitations of the technology. By following best practices, we can overcome these hurdles, save valuable time, and more quickly reap the benefits of a successful initial implementation. The following are some lessons learned from this case study that may be of some help for future implementers of library RFID systems:

- A successful RFID implementation cannot be carried out solely theoretically from an office desk.
- RFID readers and tags do not have 100 per cent accuracy in all instances. Environmental issues, the make-up of the products being tagged, and the volumes of tags to be read all impact read accuracies.
- One of the key issues for many implementations of RFID technology has been the quality of tags: both the standard inlays as well as converted labels or tags. In some instances, rates of defective tags have approached 20 per cent or more. Still more troubling are tags that work in free space under optimal conditions, but quickly lose performance when conditions worsen – these tags become inoperative where fully functional tags do not.
- Tag-reader inaccuracies are a major challenge the new technology has faced, with one of the biggest challenges being tag placement for item tagging. As RFID technology does not require direct line of sight between the reader and tags, there are a host of potential tag placement options.
- There can also be activities in daily operations that may impact on performance of the RFID installation. The purpose of a site survey is to understand requirements for the RFID system and gather information in order to specify deployment requirements. Site conditions must be taken into account when predicting and specifying performance requirements so as to be able to set realistic expectations of the system and try to avoid problems.

The other important lessons learnt during the implementation is balancing requirements and system capabilities, as some of the following requirements during the planning stage could not be realized either due to technology limitations or due to the cost of such implementations:

- Online public access catalog will be enhanced to show a graphical display of location of items. For example, a user on searching for a particular book will get the physical location information such as floor number, rack number, and shelf number along with call number. If the particular book is not in the right place or missing from the collection, the system will notify graphically, say with red color, that item is not there. All these were expected to happen dynamically in an RFID-enabled library environment.
- Patron self-checkout station was not part of the plan. It was thought that it must be possible for a user to just take the book(s) of his interest along with his member card (member card may be in his wallet) and walk through the sensor gates. The system will automatically identify items and do the checkout transactions. Users without an RFID-enabled member card cannot take books away from the library.

Suggestions

As stated previously, simply placing the equipment correctly is not sufficient to ensure maximum utilization. The following are some suggestions that may help libraries achieve optimum results during RFID implementation:

- Each library must evaluate whether RFID is the right

decision for their library and if so, determine when the timing is right. RFID technology is very much in flux. It is likely that next generation systems will be less vulnerable to misuse, will be more interoperable with other library systems, and will work better with library materials (like CDs, DVDs, and perhaps even journals) and are likely to provide features we cannot even imagine today.

- There are several issues when deciding to adopt new technology. Gathering and scanning information in order to make the right decisions are of great importance as the RFID is still relatively new to the organization. New technology must reach acceptance within the organization and funding needs to be allocated before a test can be conducted. Reluctance to properly address these issues may create problems later in the project. Another way to learn about new technology is to perform a pilot study and to learn from other implementations.
- Speak with others experienced with RFID and 'test-drive' their systems. Find out why they selected their vendor and their implementation process. Ask about how the system works in-practice. How was the customer service? Invite vendors to make a presentation. Find out equipment costs, the range of readers, and the read response times. What type of tags do they use? Find out what differentiates the vendors from one another. Develop a checklist of your requirements and compare each solution using a chart.
- A budget large enough to provide a quality degree of the required technology and technical support to go with it.
- Detailed project planning and carefully prepared implementation plans are important to the success of any RFID implementation.

- It is important that any library does not take on too much of an implementation too soon. RFID can prove challenging to adapt to and get right so it is important that it is taken one manageable step at a time if a quality outcome is desired.
- The correct type of hardware must be supported by quality software, capable of performing the tasks required of it and linking the entire complex system.
- The correct choice of supplier who will provide good after sales service as well as quality produce with security features integrated to an agreed standard.
- Top management support is one of the key recurring factors affecting the success of implementing an RFID system. Apart from setting the right strategic direction, support from top management can ensure that sufficient resources have been allotted for the development of the RFID system. Furthermore, implementation of the new system is more likely to bring about change like the abandonment of outdated processes. Resistance to change is another important issue to confront. Continuous top management support can also play a pivotal role as an agent of change to create an environment conducive to streamlining the RFID system development.
- A quality technical team trained to deal with any errors and ability to communicate well with the workforce is also a necessity so as to keep the down time of the system at minimum.
- No RFID implementation would be complete without seamless integration with ILS.
- Another important part of implementation, though sometimes performed post-implementation, is end-user education. After installation and testing, allowing the

library users to figure out the equipment on their own may be counterproductive. Experience has demonstrated that the first time the self-service equipment is used by a borrower is a critical stage. Managing this first experience must be paramount in any self-service technology introduction plan. The best way to manage the borrower's first experience is for a staff member or a trained volunteer to introduce the borrower to the system and for them to perform the first transactions together. The importance of this cannot be overstated. No amount of explanatory signage or promotional pieces can replace the assistance of a real person. The trained person is able to invite the borrower to use the self-service equipment and to remove any apprehension they might have in trying it out for themselves. The trained person can also use this occasion to emphasize the positives of the new system and to communicate information prepared by the library to allay any fears that the borrower might have about the reasons the technology has been introduced.

- With regard to placement of tags, inside of the back cover is the recommended location because it is fastest for right handed tag installers to reach. Strict consistency should be avoided and the actual height of the tag placement should be staggered at different positions to reduce the possibility that the tags of two or more books will align exactly on top of one another and cancel signals.
- During testing, it was observed that books closest to the metal separator or to the metal upright were consistently misread. These books had to be physically removed from the shelf to obtain a reading. As a result, a possible recommendation is the use of wooden shelves to enhance performance.

- With regard to placement of RFID components in the library, suitable locations for optimum results is given in Table 4.1.
- It is not necessary to commit the whole library service to RFID at the same time. For some libraries staged

Table 4.1 **Recommendation of placement of RFID units in the library**

RFID unit	Recommendation
Allocation and design of the *counter* with integrated staff station	– Close to entrance/exit – Smaller counter, as less functions are needed – No metal plates in the counter, avoid metal frames (possible influence on the RFID performance) – Distance between keyboard and reader antenna of the staff station – Sufficient space for the placement of non-processed or processed items beside the antenna
Allocation of the *self-checkout station*	– Close to the counter in order to help visitors if they have problems in handling – Consider the privacy area around (content on the screen should not be seen from waiting visitors) – Distance of at least 2 m between the self-checkout stations – Distance of at least 4 m to the sensor gate antenna – Possibility to place processed items on the side – Additional self-check stations can be placed in special sections of the library – Children and disabled people require a lower station
Allocation of the *sensor gate*	– Close to counter for direct access if an alarm is caused – Connections for video or turnstile gates if requested – Distance to door of 0.5 m – Distance to shelves of min. 4 m – No metal in vicinity – 90 cm width (or 2 × 90 cm) – Consider plants to be placed at the sides if door is wider – No floor heating system in the close area around the gate antennas (bolts cannot be placed)

Source: Kern and Nauer (2008).

implementation will be the best approach. This can be a valuable experience and provides the library with the ability to:

- monitor and assess the return on investment;
- assess customer acceptance and educational needs;
- identify impact on staff and opportunities for service adjustment;
- refine implementation processes.

The following sections provide a detailed list of key RFID components and features necessary for efficiently achieving the functionality required. This may also serve as guidelines for inviting proposal for library RFID systems.

RFID tag

- Tag must be having a read/write design, operate at 13.56 MHz and be compliant with ISO 15693 and ISO 18000 standards.
- RFID tag must have portions of memory that can be locked (for item number) and portions that can be re-programmed.
- RFID tag must not have pre-programmed number that would be rendered redundant by unique library item number.
- RFID tag must possess anti-collision capabilities allowing the reading of multiple numbers of tags simultaneously.
- RFID tags must be available to fit all standard types of library materials: books, magazines, CDs/DVDs, audio books, and videotapes, etc.
- Tags must be guaranteed for the life of the items to which they are applied.

Tagging station

- The station must be capable of programming tags by placing a tagged item on the station's antenna and then scanning the item's barcode.
- Tagging station should also be able to operate as a staff check-in/checkout station with the addition of appropriate software.

Staff check-in/checkout station

- Check-in and checkout must be capable of processing items one at a time or a stack of items simultaneously.
- When checking in items, station must display shelving location codes if they have been programmed into the tags.
- When checking in items, the system should provide both audio and visual alerts for items on reserve with information about the patron for whom the item is reserved.
- Station must provide visual and audio indication whenever a tag's security bit has been turned on or off.
- Station must be capable of turning a tag's security bit on or off independently of any other function.
- When checking in a stack of items all at one time, station must be capable of alerting user if there is a mismatch between the number of RFID tags detected and the number of items that the user has established that are to be placed on the reader at one time.
- A barcode reader must be able to operate concurrently with an RFID reader.

Patron self-service checkout station

- Station must be able to perform checkout functions and disarm tag's security bit in one process.
- Station must allow for placement of library items in any orientation.
- Station must be able to handle items comprised of magnetic media without risk of damage or erasure.
- Station must be able to process one item at a time or a stack of items.
- Station must be capable of processing multiple items in a stack (books, bound volumes, CDs, etc).
- Station must have touch screen operation.
- Patrons must have access to a detailed help screen.
- Station must be designed so that all hardware components are easily accessible and replaceable by library staff.
- Station must be designed to accommodate future hardware upgrades that may have different physical footprints, including larger monitors.
- Station should be able to use an RFID tag incorporated into patron card eliminating need for separate patron card reader (barcode, magstripe, smartcard).
- Station should be able to accommodate magnetic strip cards and smart cards.
- Station should be capable of operating in a wireless mode.

EAS security gates

- EAS gates must not damage or erase magnetic material.
- Tag detection rate must be consistent in random orientation regardless of the number of tags present in the gates' field.

- EAS gates must be able to operate in isolation of the library circulation system (e.g. no network connection to the host or server).
- EAS gates must include a people counter.
- EAS gates must possess both visual and audible alarms.
- EAS gates must be able to operate peripheral equipment such as additional alarms, door locks, and CCTV.
- EAS gates may be installed directly into flooring or into portable base plates.
- EAS gates must be capable of collecting item ID information for those items that have security bits that have not been deactivated.

Hand-held shelf and inventory management unit

- Hand-held unit must be able to scan shelves by waving a wand along the base of bookshelves without having to stop for each item.
- Hand-held unit must be capable of reading multiple items and thin items with accuracy.
- Hand-held unit must read unique item number locked in RFID tag.
- Hand-held unit must be able to download scanned item data into a computer for further importation into the library circulation system's inventory module.
- At the same time that inventory data is being collected, wand must activate the security bits on tags whose bits are in the off position if any.
- Hand-held unit must be able to search for specific item numbers by providing an audible alarm when the item has been detected.
- Hand-held wand's design must be ergonomically sound.

Patron self-service book return station

- Book return must enable returned item to be identified in the library circulation system as having been returned and update the patron account in real time.
- Book return must reactivate RFID security bits.
- Book return should be capable of printing a report for all items returned during a session and for returned items that are 'on hold.'
- Book return must have an audible alarm to alert staff when an item 'on hold' has been returned.
- Book return must have the option of attaching a receipt printer that prints individual tickets for items on hold.
- Book return must be designed so as to interface with automated sorting equipment.
- Book return must be able to process library materials without danger of damaging or erasing magnetic media (e.g. video cassettes).
- Book return should be capable of operating in a wireless mode.

Conclusion

Deciding to implement RFID raises a new set of tasks such as setting goals, evaluating other installations, selecting a vendor, planning conversion, managing the actual conversion, and assessing the results. Many varieties of RFID exist. Each needs to be verified independently. The technical and economic differences among the varieties dictate that decisions regarding the choice of systems and cooperation of users, management, system integrators, and other solution providers hold the key to successful implementation of the technology.

RFID is not yet a plug-and-play commodity technology. Some providers will take a one-solution-fits-all approach, which invites complications and problems. Even when the optimum design is selected, it may need a custom design specific to the application to achieve optimum performance. Tradeoffs often need to be evaluated. RFID under-performs in some applications because of a non-optimized solution approach. A basic understanding of RFID, its varieties, and custom design tools is important when evaluating its potential use. Too often, the underlying engineering and physics are not understood, minimal training is provided, and expectations are unrealistic. Product selection mistakes early on can cause tremendous increases in costs later in the project. Libraries must be careful to keep their focus on the benefits that RFID brings to their organizations.

Though the RFID landscape can be confusing as vendors claim they can do it all, the claims are often conflicting and difficult to verify. End users looking to adopt RFID today have a big advantage over their predecessors because the technology, software, and know-how are improving. Generally, RFID will present benefits to the adopter in most circumstances, but to maximize benefits one must address the variables. Those variables include the workforce, workspace, electronic interferences, monetary restrictions, previous experience (internal and external), etc. Only after considering all the variables a decision can be made smartly. Optimally implementing RFID is challenging and requires both creative vision and careful execution. But nothing worthwhile is not challenging in life and in business. And, when properly approached, that small little RFID tag can just be a catalyst that takes the library into a state of users' bliss. The adoption of RFID technology for self-services without doubt will make the libraries significantly more user-friendly and efficient.

RFID issues

Introduction

Change is an issue that has received emphasis, as it must be understood that any modification of an existing organizational environment does not simply require a process focusing on locating a new improved piece of technology with a good reputation and placing it into the current system. In actual fact, a good deal of thought must be taken as to whether the technology would fit the organization's environment and way of working, the effects that such a dramatic change may have and if the current employees really are ready for it (Nightingale, 2009).

Information and communication technologies are playing an increasingly important role in the implementation of value-added processes that overarch several steps. As organizations invest in technological innovations to gain competitive advantage in the marketplace, they need to realize that although technology is an enabler, the adoption of new technology could pose a complex challenge, especially with the rapid pace of technological change. The most recent technology in which organizations are investing to gain competitive advantage is radio frequency identification (RFID). RFID as a symbol of a new paradigm eliminates the separation between the physical goods and the information corresponding to it. This technology has the potential of impacting an organization's business profoundly. Many organizations have adopted this technology and others are in the process of doing so. Although the initial drive for RFID adoption was supply chain management, today many organizations are involved in implementing RFID to solve business problems (Borriello, 2005). RFID is a powerful technology that holds great benefits and promises to revolutionize the item-level tracking.

Emerging technologies that relieve the library staff from time-consuming operations have always been of interest to libraries. In today's fast moving world, existing systems for checking items in and out of libraries increasingly seem slow, labor intensive, inflexible, and prone to inaccuracy. Libraries are fast growing adopters of RFID as the technology promises to relieve staff from repetitive circulation jobs, speed patron self-check, reduce pilferage, and provide accurate and timely inventory management. RFID-based systems improve convenience and flexibility for both the users and the library. Quicker and easier to operate, these systems enable improved inventory control, minimize losses via internal misplacement and theft, reduce manual work for library staff, and deliver better read rates and accuracy than traditional manual handling. Another function made possible with RFID systems is fast and easy location of books. This can be achieved either by placing antennas on shelves to define their contents or with a handheld proximity reader, and enables quick inventorying and the location of misplaced items. Despite these promising applications of RFID in libraries, a number of challenges have hampered the adoption of RFID. It has been suggested that library RFID applications may be the first major deployment of item-level tagging. The library provides the panorama to study the operational issues of RFID tagging at the item level. While the literature has discussed a number of perceived and expected problems with RFID in the supply chain, it is sparse on performance and reliability of RFID library systems. The implementation issues highlighted in the supply chain include failed or erroneous reads, handling of large amounts of data generated by the tags, lack of mature standards, collision, and failure in the presence of metal- and liquid-based products. Investigation of RFID in the library environment is important, as it is decidedly different from a distribution or manufacturing environment. The major RFID issues in libraries can be broken down into technology issues, social issues, and economic issues. Let us look at them in detail.

Technology issues

Technological uncertainty reflects the risk that RFID may fail to deliver the expected benefits within the anticipated time frame or that it may be superseded before the benefits are realized. Technology risks and challenges with RFID originate from many sources. One primary issue is the current reliability of the technology itself. RFID systems face many technical challenges and obstacles to practical adoption. A major hurdle is simply getting RFID systems to work in real-world environments. Systems that work perfectly in a lab setting may encounter problems when faced with environmental noise, interference, or human elements. A great number of variables affect the characteristics of all RFID systems, including (NISO, 2007):

- width between security gates;
- number of items simultaneously exiting the library;
- material of which the items are made;
- size of the RFID tags;

- tuning of the antennas on the RFID tags;
- orientation of the tags in the portal;
- tags' relative positions to each other.

There are a host of other technical problems that affect RFID technology, which can be grouped as hardware, software, integration, and standard issues.

Hardware issues

Not all the applications of technology and product are perfect in every way, especially RFID, which utilizes radio frequencies to transmit signals. Radio frequency interference problems and reading difficulties concerning read ranges and read rates are most inherent to RFID technology. The interference and detected accuracy in the transmitted process influences actual efficiency of the application. The interference with metal and fog, the distance between reader and tag, and the antenna direction are factors leading to loss of detection. Generally the effective read range of an RFID system is dependent on many different factors including:

- tag-reader frequency;
- tag energy efficiency;
- alignment of the tag's antenna with respect to the reader's antenna;
- coupling between the tag and the reader;
- antenna design as well as the shape, size, and quality of both the RFID tag and reader;
- sensitivity of the reader;
- transmitting power generated by the reader;
- environment (outdoor and indoor);

- susceptibility to noise and interference from other radio devices.

The problem with read rates is simple, in that tags passing by a reader may not register, may register to an adjacent reader, or may even register incorrect information. Tag and reader manufacturers have been working diligently to resolve these problems, despite the reality that variability in environmental conditions is almost infinite. Attempting to read several tags at a time may result in signal collision and ultimately to data loss. To prevent this, anti-collision algorithms can be applied at an extra cost. The development of these methods, aimed at reducing overall read time and maximizing the number of tags simultaneously read, still goes on (Burdet, 2004).

The other hardware problems include RFID reader collision and RFID tag collision. RFID reader collision occurs when there is an overlap in the coverage area of multiple RFID readers. Because RFID tags are designed to only be able to transmit to one reader at a time, this can cause several problems. One is that the signals from overlapping readers can interfere and prevent the item from being scanned. Another is that the signal convergence can lead to the same tag being read more than once, leading to duplicate and inaccurate scanning results.

Not only do metal, mist, and distance influence the read/write efficiency of RFID, incorrect positioning of antennas also causes failure even if the tag and the reader are very close together. Besides, too close a distance among tags may produce interference between tags or erroneous access. For example, if one patron is taking books to initiate the loan process under self-check-in/checkout equipment (RFID reader), and another patron is standing too close to the first patron, the reader doing the self-check-in/checkout may detect the tags of books that are held by the wrong patron (Yu, 2007).

Tags may be damaged during usage. A wide range of application challenges can be answered by the multitude of suitable tags, yet none of them is completely invulnerable and the causes of damage may vary from type to type. The result is a read failure, which is, in many cases difficult to detect, as is the fact of the damage itself for a hidden tag.

Adverse conditions of the environment and improper placement may corrupt reading. As mentioned before, absorption, ambient reflection of the signal and external signal sources (such as security systems, cordless phones, and barcode scanners) may introduce read errors. Similarly, improper orientation of tags may impair reading efficiency, as most antennas used in tags are direction-sensitive. Manufacturing of tags is not yet 100 per cent failure-free today; about 20–30 per cent of tags used in early RFID pilots have been defective (Lewis, 2004).

RFID tags are not hidden since these tags are typically affixed to the inside back cover and are exposed to vandalism. Some interference from metallic material in book covers, CDs, and DVDs does occur and tag reading may not be correct. RFID tags are easy to shield from readers by ordinary aluminum foils and may not serve the purpose of being anti-theft tools.

RFID reader range depends on its power and antenna size. With malfunctioning RFID readers, the difficulty is in knowing when malfunctions occur.

Maintenance of security gates: A disadvantage of electronic security systems located at exit points in the library is that they create a false sense of security, and detection can also be overcome by power failures, or by electrical or electronic faults. Their typical success is preventing the absent-minded patron from taking books out of the library, or the novice thief. No electronic book theft detection system is foolproof and no security system can eliminate book theft (Mansfield, 2009).

Software issues

Any software solution that is developed today will at best be a stop-gap arrangement as the technology itself is evolving. New technologies are being introduced with different RF physics, transmission schemes, supporting different frequency bands, new protocols, new standards, multi-protocol support, changing governmental regulations, etc. Thus, any software solution developed now will have to evolve with the RFID technology should be extendible and adaptable with minimum disruption to the deployed infrastructure (Prabhu et al., 2006).

Most RFID systems use the SIP2 or NCIP protocol to communicate with the library management system (LMS). This means that the currently used software for the library's operations may not meet the requirements for the RFID system. In such cases, the library's current software will need to be modified to include the SIP2/NCIP standard or the library will have to switch to some other software package that does support the standard and re-enter hundreds to millions of items, depending on the size of the library. Also, the middleware based on SIP2 or NCIP for integration of RFID systems into the existing LMS available today is still not tailored to the needs. LMS-RFID integration is very important for everyone involved in RFID in libraries. This includes LMS suppliers, RFID suppliers, and of course libraries. When an RFID tag only contains a unique number (accession number), which identifies that copy of the book, that is the only piece of data which can be communicated between the RFID system and the LMS system. This data would normally come from the LMS. The interface between the RFID and LMS would have to be able to exchange all this information (Edwards and Fortune, 2008).

System integration issues

Like most information technology systems, RFID systems still require practical expertise to install, configure, and manage. End-users should expect to experience mundane technical complications that arise while implementing RFID. Despite marketing claims to the contrary, RFID is not a 'magic bullet' that is simple to implement out of the box. In order to function appropriately, the RFID system must be integrated with other information systems. Internal integration requires that an organization's existing facilities, equipment, hardware, and software be examined in light of the proposed RFID system, and then procedures put in place to insure a smooth implementation of the new RFID system (Viehland and Wong, 2007).

Standards issues

Technological standards are one of the current issues that the RFID industry until now has failed to address and are a barrier to its popularity over more established technologies. Although using useful standard elements from similar technologies there is unfortunately a lack of a 100 per cent agreed upon standard dedicated to the technology, meaning that any RFID's implemented in the current time period run the risk of becoming rapidly out of date and furthermore incompatible with any additional RFID plans that an organization may choose to implement in the future. The tags used by most vendors of library RFID are not compatible even when they conform to the same standards because the current standards only seek electronic compatibility between tags and readers. The pattern of encoding information and the software that processes the information differs from vendor to vendor; therefore, a change from one vendor's system to another would require modifying all the software. A unified, globally interoperable RFID standard is ideal to realize the full benefits of RFID applications. Lacking technological standards and the current uncertainty of technological standards are major barriers to RFID implementation. Lack of uniformity in RFID technology and standards keep the costs high, and standards are the key to the proliferation of RFID technology. The establishment of a standard will force the cost to drop since RFID product suppliers can all produce compatible chips, readers, associated hardware, and software.

Social issues

Privacy

One issue often linked with RFID technology is the general consensus that it impedes the privacy of those affected; its monitoring capabilities can potentially obtain data that patrons may not wish the library to posses, their location for example. Although a lot of these privacy issues are claimed to be over hyped, it is still recommended that any academic library wishing to carry out an RFID project creates a quality privacy policy and reviews it continually (Nightingale, 2009). The privacy issue is also a subject of debate in the application of RFID technology. All patron activities such as reading, browsing, and action behavior will be detected by readers that are installed in libraries. These observing operations are similar to surveillance and may be a privacy problem for the patron. Such issues and concerns about privacy may prevent the application of active services in libraries.

Economical issues

Costs

The cost of the RFID system and its integration to existing information systems is a major challenge in the global roll-out of the technology. RFID systems require expenditures for tags, readers, hardware, software, and system maintenance. Prices, in particular those associated with the smart tags, are yet to settle at a reasonable level, certainly in comparison with the more traditional library system methods such as barcode technology (Nightingale, 2009).

The cost is one of the major factors influencing acceptance of RFID, although the production costs of RFID have reduced. This is especially the case with item-level tagging. For library collections with over 100,000 items, it will be hugely expensive to implement RFID completely. Plus, the costs for the reader, peripheral equipment, and application software will be even more of a burden. This will stretch the budget, and the schedule for implementation of an RFID solution will be hard because libraries hold enormous collections (Yu, 2007). Description of costs of various RFID components are given below:

- *RFID tags*. These are usually the most important costs associated with RFID technology. They are fairly expensive and the cost depends on the form factor, memory capacity, read or read-write capability, active or passive configurations and range. The cost of the tags can further be broken down into three major elements: the silicon chip, the copper antenna, and the process of joining the two. To be economically viable, the cost of the tags will have to be brought down.

- *RFID readers.* Reader costs vary depending upon the features such as network enabled or wireless readers. When more readers are planned, the price of the readers could also be a critical factor.
- *Middleware costs.* Middleware cost is another component that adds up the cost in addition to the cost of the library management system.
- *Integration costs.* No matter what the middleware chosen, integration costs will be a large factor in RFID implementation. Some systems may need to be replaced altogether if they are not capable of taking advantage of incrementally large volumes of real-time data.
- *Installation costs.* The cost of the installation of the RFID equipments, affixing RFID tags on items, and programming.
- *Maintenance costs.* Factors influencing maintenance include hardware upgrades, replacement of failed or damaged equipment, middleware upgrades or licensing fees.
- *Costs of the training of personnel.* When introducing new processes and technology, it is necessary to train the personnel.

Return on investments (ROI)

Return on investment is a measure used to evaluate the efficiency of an investment or to compare the efficiencies of number of investments. To calculate the ROI, the benefit of the investment must be divided by the cost of investment and the result is expressed as a percentage or ratio. In the commercial world, all things are measured by 'return on investment' or ROI. The cost of the new technology is compared to the increase in profits. When libraries measure their success, profit is not part of the equation as with other institutions that provide services, such as educational institutions and research establishments. Libraries 'spend' their ROI on new services or on beefing up existing services. They also spend their ROI to respond to budget cuts or to loss of buying power when budgets do not keep up with inflation. This makes it hard to demonstrate that an investment in technology is worth the cost.

Libraries do have to justify their technology budgets to their boards or their institutional oversight committee. However, some amount of the expenditure on technology is not a matter of a measurable return but is simply the cost of continuing to be a player in a high-technology information society. Any library that fails to keep abreast of technological change will be quickly seen as irrelevant in the larger information context. Libraries do not usually perform rigorous analyses to justify their purchases of equipment. There is an assumption in the profession that libraries cannot calculate a return on investment because our services are intangible and cannot be measured (Coyle, 2005).

The following are factors to be considered in assessing ROI (Smart, 2004):

- time, labor, and materials costs for processing new materials;

- time and labor associated with checking-in, sorting, and shelving materials;
- percentage of staff time spent on the above circulation tasks;
- percentage of staff time spent training and managing part-time workers/volunteers performing the tasks listed above;
- number and value of lost items for a specified time period;
- time spent searching for lost items;
- time spent doing inventory and the amount of inventory completed;
- time and labor spent in shelf reading and maintenance;
- percentage of circulation currently being done by self-check;
- average wait at circulation desk during a busy period;
- number of staff at the circulation desk during a busy period;
- number of requests caused by mis-shelved materials;
- ongoing equipment maintenance costs.

RFID in the digital environment

While the role of the library continues to evolve with advancement of the internet and other information technologies, the fact remains that the library is still the premier warehouse for the written word. However, in many libraries, the requirement to hold printed materials is declining with time, as more efficient delivery methods for information evolve. While this is true within the public library sector, it is especially being felt within academic libraries. The application of RFID within libraries is tied very closely to physical assets and their management. Whether it is collection management, security, or circulation, the physical library assets are what are being considered in any discussion of RFID. If online delivery of information is progressively displacing the need for physical assets, academic libraries might question the benefit of investing in technologies designed to manage items for which they have a reducing need. This is not to suggest that libraries will cease to carry books, but rather to acknowledge that the benefits delivered by physical asset management systems such as RFID may gradually be eroded if current academic library trends continue (Lewis, 2006).

Discussion

One of the common concerns in implementing RFID today is the rapid obsolescence of the technology, especially in view of the investment cost. Technology is continuously evolving and new protocol standards, faster and more fault-tolerant readers quickly outdate their predecessors. Failure to detect tags that are present in the read range of a reader can be due to a variety of causes including collisions on the air interface, tag detuning, tag misalignment, and metal and water in the vicinity of the RFID system. It is imperative that RFID systems eventually approach 100 per cent reliability. Costs remain the largest impediment for the widespread adoption of RFID. RFID is not cost effective if seen as a transactional technology alone, since cost savings at an operational level will not be sufficient to justify the investment. Significant effort would be required to obtain the necessary capital funds to implement RFID solution, and the operating costs required to make it sustainable.

A key to reducing costs is to increase the volume of demand; this can be facilitated if a unified global RFID standard exists. A universal RFID chip, which can be used by different applications in different countries, is an ideal to pursue; this ideal universal RFID chip would have a frequency-independent capability and would be integrated with read/write memory. Furthermore, for an RFID chip to satisfy the requirements of multiple applications, a minimum set of commands must be designed so that it can be customized by software for a variety of applications (Wu et al., 2006).

Neither EM nor RFID are 'security systems;' they are 'theft deterrent' solutions. A real library security system that would guarantee protection is called a closed stack. We need the best tags and the best gates on the market if part of the mission is protection. Most of the RFID gate systems on the market are 2D, which means they will not detect an item in all positions through the gates. That is a 33 per cent performance sacrifice just for the gates. If protection is a metric of investment, then 3D gates are essential but that add up the costs.

While the challenges to worldwide adoption of RFID are significant, they are not insurmountable. As with many other emerging new technologies, it will simply take time for reality to meet ends with the ideals. As more efforts are put in for RFID research and development, more pilots will be run and more data will be collected; these can be used to solve problems that stand in the way of widespread adoption. This is an ongoing process.

But for now, it has to be made clear that though the RFID technology promises a number of fancy characteristics, its real application is not straightforward. Many people think RFID is magic – that it just works. RFID is a costly system that offers some great advantages over barcodes. Even though the technology still hasn't matured, the opportunity for immediate and dramatic improvements in customer service and productivity outweigh the negatives. Improvements in standardization have already occurred, with several additional standards under development. Most large system providers have mainly resolved the problems with interoperability. Improvements in tag life expectancy and durability have improved substantially in the past few years. The benefits of RFID technology cannot be ignored in today's tight staffing and funding climate. RFID technology is maturing and continues to see tremendous innovation. Today, RFID systems have become broader, deeper, and cheaper. Readers are using less power and are operating faster and at longer distances and with more ability to handle interference. For more widespread application, RFID technology must overcome some of the challenges that slowed its initial development. Organizations that are implementing RFID today need to be aware of the upcoming technological developments. They need to prepare their organizations for the next generation of technology. Companies have to get away from thinking of RFID as a cost-cutting tool and look at its broader strategic role and explore its potential. RFID technology needs to be treated as part of a suite of technologies helping to enable different business processes and become a mainstream part of information technology systems.

Conclusion

RFID has come a long way but arguably still has some way to go before it lives up to the original hype. With all these promises, it is not easy to separate truth from hype. Technical reality changes constantly, and within hype might be found a note of reality. We must walk the fine line between believing the benefits of the technology touted by vendors, and knowing how it will solve real-world business issues. Vendors with heavily invested RFID resources obviously promote the existing, future, and even dreamed-up advantages of RFID, but they seldom shed light on obvious and much lower-costing alternatives. The key to finding the optimal use for RFID is to begin at the beginning and first consider your organization's strategic objectives. RFID must be applied in a balanced way that takes into consideration strategic objectives, without being carried away by the hype of RFID (Brooke, 2005). RFID offers significant potential benefits, but one must consider the entire picture of an organization in order to see the greatest benefit. Additionally, do not mistake the best solution to be the one with the highest price tag. Expensive solutions are not the answer. Another way RFID will change is in its architecture. Today, the common model for RFID employs a tag that communicates with a reader, but there is no tag-to-tag communication. Having tags that talk to each other creates some interesting possibilities. When all is said and done, there is only one certainty: RFID will change.

6

Future development of RFID

RFID technology combines two important features: automatic identification (the 'ID' part) on the one hand and data transmission via radio communication (the 'RF' part) on the other hand. To date, no alternative technology really combines these two parts. RFID, its application, standardization, and innovation are constantly changing. Though its adoption is still relatively new and hence there are many features of the technology that are not well understood by the general populace, many developments enabling widespread RFID adoption are already realities (Chandak, 2008):

- Efforts are on standards to allow end users to multi-source RFID solutions and components out of the market.
- RFID integrated circuits (IC) are mature, and multiple players are offering ICs for RFID frequencies (LF, HF, and UHF).
- Hardware (readers, handhelds, and printers) are now mature for low frequency (LF) and high frequency (HF), and are quickly maturing for ultra high frequency (UHF) technology.
- Technology know-how is available to optimize the performance of the RFID solution for the challenges/applications of the customer.

- Competence (systems know-how, business calculations, and the system integration) necessary for creating ROI is also in place.
- Technology discussions have been successful on topics such as which frequency should be used for which application.

RFID continues to evolve with two key goals of reducing cost and adding functionality for end-users. Developments in RFID technology continue to yield larger memory capacities, wider reading ranges, and faster processing. It is highly unlikely that the technology will ultimately replace barcode – even with the inevitable reduction in raw materials coupled with economies of scale, the integrated circuit in an RF tag will never be as cost-effective as a barcode label. However, RFID will continue to grow in its established niches where barcode or other optical technologies aren't effective. If some standards commonality is achieved, whereby RFID equipment from different manufacturers can be used interchangeably, the market will very likely grow exponentially. The main technological developments include (OECD, 2007):

- *Improvement of specialized tags*. Currently, tags for difficult environments, such as metallic environments, and conditions such as high temperature and rough weather conditions, are available. These specialized tags are being further developed and improved. Overall, the variety of tags is supposed to increase so that there will be multiple highly specialized tags for specific environments and conditions.
- *Improved battery supply*. An important obstacle for the use of active technology is the limited lifespan of an active tag due to factors such as short-lived batteries, the design of wake-up protocols, and whether tags are read

continuously or at pre-set intervals. In the next few years, further improvements are very likely.

- *Improved read-ranges.* Further improvements are also expected in the field of current read ranges for particular frequency ranges.
- *Use of new materials for RFID tags.* In order to further significantly decrease costs of RFID tags, new materials are being tested. It is hoped that polymer-based chips will replace RFID tags with silicon-based chips currently in use. By using polymer-based chips, costs could be significantly reduced and tags could serve as a low-cost means for a broad use of RFID at the item level. Prototypes are already developed and different polymers are being tested.

RFID systems are part of a bigger picture and are potentially a key stepping-stone in the development towards the vision of ubiquitous computing. In the ubiquitous or pervasive computer vision there will be a multitude of computationally capable, small – sometimes invisible to the human eye – devices that will be scattered throughout our environments, operating silently and largely unseen as they go about their individual tasks to support daily activities. In a step-change that will be orders of magnitude greater when compared to today's computing power, a bewildering population of heterogeneous sensors, computers, and actuators will be operating. Often, these devices will operate with self-awareness (being 'conscious' e.g. of their physical location and their immediate surroundings) and be widely networked together. In order to realize this vision a comprehensive jigsaw of technological 'pieces' needs to come together and converge as technology develops over the next few years. The RFID piece of the jigsaw is the ability for individual (physical) items to be able to identify themselves to the network. In an information-rich, digitally connected world, where much of the knowledge and tools that we make use of are outside our heads there will be a need to develop new communication 'senses' that allow us to manage and make use of the enormous amount of information we will be confronted by. This will lead to the development and adoption of new and different types of human–computer interfaces and different ways of communicating with technology (Ward and Kranenburg, 2006). RFID by means of chips is not the only means of advanced technological identification. Alternatives are being found that include chipless RFID systems, biometric identification techniques, and DNA-markers (Lieshout et al., 2007).

With the anticipated merging of the two most prominent RFID standards, i.e. EPC and ISO, it is projected that the unified standard will be the catalyst for the future development of RFID. It will not only enhance the interoperability of products and systems but also facilitate global data synchronization and sharing. The emerging RFID standard and the anticipated price drop in RFID tags and devices will be the two strong driving forces for the wider adoption of RFID. Over the next 15 years, the money spent on RFID tags will peak then rapidly fall. Perhaps RFID repeats the history again of barcodes – companies today do not pay per barcode – they are printed usually during the package production process and are virtually free (Lee, 2008).

RFID printing is gaining grounds, and in the future, several products printed with tags for the purpose of authenticity and identification will emerge. There are continuing discussions within the library community about compatibility with the supply chain of the publishing industry. One school of thought suggests that publishers or distributors may choose to employ RFID tags at the point of manufacture, in order to better track their shipments and also to provide stock management and item-level security at retail outlets. If publishers more broadly should choose to employ UHF systems in common with other supply chain systems, libraries may be advantaged if they are able to use the same RFID tag that is provided with each item (Butters, 2008a, 2008b). If every book had incorporated into it during production a tag, this could contain data such as the ISBN plus copy number, which together would make a unique identifier. This could be a write-once tag, its data would never change. This might be used by library systems and equated to the library's item identification number by means of a table held in the database of the automated library system. The publisher, wholesale distribution, retail distribution, and the library could therefore use the same tag successively. It can be seen that a library would benefit more from RFID being incorporated into a book than the book trade. RFID as envisaged in this scenario would be most effective if it were incorporated into the majority of books.

Rarely in our history have we faced technological challenges such as the ones that we face today. And it is our duty to mobilize our knowledge and cultural values to address them. RFID technology as we know it today will evolve and reach unprecedented levels of functionality, of memory storage, and processing capabilities. In the near future, we will see the breakdown of the boundary between cyberspace and real space. The worlds of data and things will merge so that the virtual world of the web will be rendered physical as we move towards – what computer scientists have called 'the Internet of Things.' We're heading for a world in which billions of networked objects and sensors will report their location, identity, and history. Such a phenomenal development, hard to imagine not long ago, should open up tremendous opportunities for both economic prosperity and the quality of life of citizens. For instance, over the next 50 years, the innovative marriage of RFID tags, sensors, Galileo, 4G networks, Wi-Fi, and artificial intelligence will create an 'intelligent infrastructure' that has the potential to dramatically reduce congestion and pollution, and enhance security, passenger monitoring, and comfort (Reding, 2006).

References

3M. Online at: *http://www.3m.com.*

AIM. *Radio Frequency Identification – RFID: A Basic Primer.* December 2008. Online at: *http://www.8051projects.info/files/RFIDPrimer.pdf.*

Al-Muhamed, M.S. (2007) *Radio Frequency Identification (RFID) for Automating Business Processes.* December 2008. Online at: *http://www.gisdevelopment.net/proceedings/mest/2007/Papers/day1/P12.pdf.*

Apte, U.M. et al. (2006) *What Is the Right RFID for Your Process?.* December 2008. Online at: *http://www.dtic.mil/cgi-bin/GetTRDoc?AD=ADA498749&Location=U2&doc=GetTRDoc.pdf.*

Asif, Z. and Mandviwalla, M. (2005) 'Integrating the supply chain with RFID: a technical and business analysis.' *Communications of the Association of the Information Systems, Vol. 15.* December 2008. Online at: *http://www.bauer.uh.edu/rfid/Spring2006/RFID_Tutorial.pdf.*

Banks, J. and Thompson, L.G. (2008) *Understanding RFID Part 6: RFID Readers.* December 2008. Online at: *http://www.rfidnews.org/2008/04/21/understanding-rfid-part-6-rfid-readers.*

Barbosa, A.C., Porto, F., and Melo, R.N. (2002) 'Configurable data integration middleware system,' *Journal of the Brazilian Computer Society,* 8(1): 12–9.

Bibliotheca. Online at: *http://www.bibliotheca-rfid.com.*

Boone, T. and Studwell, B. (2006) *Planning for RFID and Related Privacy Issues*. December 2008. Online at: *http://www.aallnet.org/products/pub_sp0611/pub_sp0611_Barcodes.pdf*.

Boss, R.W. (2007) *RFID Technology for Libraries*. December 2008. Online at: *http://staging.ala.org/ala/mgrps/divs/pla/plapublications/platechnotes/RFID-2007.pdf*.

Borriello, G. (2005) 'RFID: tagging the world,' *Communications of the ACM*, 48(9): 34–7.

Brooke, M. (2005) *Common Mistakes, Uncommon Best Practices*. December 2008. Online at: *http://www.rfidjournal.com/article/articleview/1483/1/128/*.

Brown-Syed, C. (1998) 'Library and archival security in the next century,' *Library and Archival Security*, 14(2): 1–4.

Bumbak, M. (2005) *RFID and Security: Analysis of Potential RFID Security Problems in Supply Chains and Ways to Avoid Them*. December 2008. Online at: *http://dexus.amaze.nl/~rfidweblog/docs/Bumbak_thesis_RFID%20and%20security.pdf*.

Burdet, L.A. (2004) *RFID Multiple Access Methods*. December 2008. Online at: *http://www.vs.inf.ethz.ch/edu/SS2004/DS/reports/06_rfid-mac_report.pdf*.

Butters, A. (2006) *Radio Frequency Identification (RFID): An Introduction to Library Professionals*. December 2008. Online at: *http://www.sybis.com.au/GeneratedItems/RFID%20Whitepaper.pdf*.

Butters, A. (2008a) 'RFID in Australian academic libraries: exploring the barriers to implementation,' *Australian Academic and Research Libraries*, 39(3): 198–206.

Butters, A. (2008b) *New RFID Technologies and Standards – What Does it All Mean for Your Library?*. November 2009. Online at: *http://www.valaconf.org.au/vala2008/papers2008/66_Butters_Final.pdf*.

Cadoo, S. (2004) *RFID Use within Libraries: An Australian Perspective*. December 2008. Online at: *http://conferences.alia.org.au/newlibrarian2004/zobjects/presymppapers/Cadoowebsitepaperfinal.pdf*.

Chandak, A. (2008) *RFID Apps Climb Maturity Curve*. November 2009. Online at: *http://www.ciol.com/Semicon/Special-Report/Feature/RFID-apps-climb-maturity-curve/111208113603/0/*.

Chao, C.-C. et al. (2005) 'Determining technology trends and forecasts of RFID by a historical review and bibliometric analysis from 1991 to 2005,' *Technovision*, 27(5): 268–79.

Checkpoint Systems. Online at: *http://www.checkpointsystems.com/*.

Ching, S.H. and Tai, A. (2009) 'HF RFID versus UHF RFID — technology for library service transformation at City University of Hong Kong,' *Journal of Academic Librarianship*, 35(4): 347–59.

Coyle, K. (2005) 'Management of RFID in libraries,' *Journal of Academic Librarianship*, 31(5): 486–9.

Curtin, J., et al. (2007) 'Making the "MOST" out of RFID technology: a research agenda for the study of the adoption, usage and impact of RFID,' *Information Technology and Management*, 8(2): 87–110.

Das, R. (2006) *RFID in Action: Summary*. Cambridge: IDTechEx.

Das, R. and Harrop, P. (2006) *RFID Forecasts, Players and Opportunities 2006–2016: Summary*. Cambridge: IDTechEx.

Das, R. and Harrop, P. (2009) *RFID Forecasts, Players and Opportunities 2009–2019: Summary*. Cambridge: IDTechEx.

Dawes, T.A. (2005) 'Is RFID right for your library?,' *Journal of Access Services*, 2(4): 7–13.

Dempsey, M. (2004) 'Five steps to successful RFID deployment,' *Ascet*, 6: 132–4. December 2008. Online at: *http://www.ascet.com/content/PDF/ASC6_wp_dempsey.pdf*.

Dorman, D. (2003) *Implementing RFID Technology in a Consortial Environment Using a Shared Library Management System*. December 2008. Online at: *http://www.lincolntrail.info/RFIDInConsortialEnvironment.html*.

Edwards, S. and Fortune, M. (2008) *Guide to RFID in Libraries*. November 2009. Online at: *http://www.bic.org.uk/files/pdfs/090109%20library%20guide%20final%20rev.pdf*.

Erwin, E. and Kern, C. (2003) 'Radio frequency identification for security and media circulation in libraries,' *Library and Archival Security*, 18(2): 23–38.

Glover, B. and Bhatt, H. (2006) *RFID Essentials*. Sebastopol, CA: O'Reilly.

Gragg, J. (2003) *Emergence of RFID Technology in Modern Society*. December 2008. Online at: *http://www.thaieei.com/embedded/pdf/RFID/40017.pdf*.

Gupta, M. (2008) *RFID in Libraries Guide*. December 2008. Online at: *http://www.streetdirectory.com/travel_guide/print_article.php?articleId=137379*.

Guy, F. (2007) 'Technological advances and information education 1982–2007: some perspectives,' *Education for Information*, 25(1): 63–72.

Heng, S. (2004) 'Rising stars in information and communication technology,' *Economics* (46). December 2008. Online at: *http://129.3.20.41/eps/dev/papers/0411/0411015.pdf*.

Heng, S. (2009) 'RFID chips: enabling the efficient exchange of information,' *Economics* (69). November 2009.

Online at: *http://www.epcglobalinc.org/home/Deutsche_Bank_RFID.pdf*.

Hodges, S. and McFarlane, D. (2005) *Radio Frequency Identification: Technology, Applications and Impact*. December 2008. Online at: *http://www.autoidlabs.org/uploads/media/AUTOIDLABS-WP-HARDWARE-016.pdf*.

Karygiannis, T. et al. (2007) *Guidelines for Securing Radio Frequency Identification (RFID) Systems*. December 2008. Online at: *http://csrc.nist.gov/publications/nistpubs/800-98/SP800-98_RFID-2007.pdf*.

Kern, C. and Nauer, M. (2008) *Implementing RFID in Libraries for Process Automation – Experiences from over Twenty Current Installations*. December 2008. Online at: *http://www.bibliothecarfid.com/files/pdf/whitepapers/biblio_ExperiencesFromImplementing.pdf*.

Lahiri, S. (2005) *RFID: A Technology Overview*. December 2008. Online at: *http://www.ibmpressbooks.com/articles/printerfriendly.asp?p=413662*.

Landt, J. (2001) *Shrouds of Time: History of RFID*. December 2008. Online at: *http://www.transcore.com/pdf/AIM%20shrouds_of_time.pdf*.

Lee, S. (2008) *Printed Electronics: the Future of RFID?* November 2009. Online at: *http://www.printedelectronicsworld.com/pages/printarticles.asp?articleids=811*.

Lewis, D. (2006) *A Model for Academic Libraries 2005–2025*. December 2008. Online at: *http://mulibraries.pbworks.com/f/C__DOCUME~1_SWINDE~1_LOCALS~1_Temp_plugtmp-12_A+Model+Academic+Libraries+2005+to+2025.pdf*.

Lewis, S. (2004) *A Basic Introduction to RFID Technology and Its Use in the Supply Chain*. December 2008. Online at: *http://www.idii.com/wp/LaranRFID.pdf*.

Liard, M. and Carlaw, S. (2009) *RFID Annual Market Overview – Excerpt*. November 2009. Online at: *http://www.upmrfid.com/upm/internet/upm_rfid.nsf/images/ABIMarketOverview.pdf/$FILE/ABIMarketOverview.pdf.*

Lieshout, M. et al. (2007) 'RFID technologies: emerging issues, challenges and policy options,' JRC scientific and technical reports, Institute for Prospective Technological Studies. December 2008. Online at: *http://ftp.jrc.es/EURdoc/eur22770en.pdf.*

Mansfield, D. (2009) 'Reducing book theft at university libraries,' *Library and Information Research*, 33(103): 10–15.

Michalek, I. and Vaculik, J. (2008) *History and Evolution of RFID Technology*. July 2009. Online at: *http://ks.utc.sk/casopis/pdf/III2008/michalek_vaculik.pdf.*

Minami, T. (2009) 'A design for library marketing system and its possible applications.' In *Lecture Notes in Artificial Intelligence*, Vol. 5465, edited by D. Richards and B.H. Kang. Springer, pp. 183–97.

Molnar, D. and Wagner, D. (2004) *Privacy and Security in Library RFID: Issues, Practices and Architectures.* December 2008. Online at: *http://www.cs.berkeley.edu/~dmolnar/library.pdf.*

Ngai, E.W.T. et al. (2008) 'RFID research: an academic literature review (1995–2005) and future research directions,' *International Journal of Production Economics*, 112(2): 510–20.

Nightingale, G. (2009). *Reality of RFID Technologies in Academic Libraries – Using an Individual Case Study and Nationwide Survey to Investigate RFIDs Potential.* November 2009. Online at: *port.ac.uk/636/01/2009NightingaleG.pdf.*

National Information of Standards Organization (NISO) (2007) *RFID in US Libraries*. December 2008. Online at: *http://www.niso.org/publications/rp/RP-6-2008.pdf.*

National Institute of Standards and Technology (NIST) (2007) *Guidelines for Securing Radio Frequency Identification (RFID) Systems*. December 2008. Online at: *http://csrc.nist.gov/publications/nistpubs/800-98/SP800-98_RFID-2007.pdf.*

Organisation for Economic Cooperation and Development (OECD) (2007) *RFID Implementation in Germany: Challenges and Benefits*. December 2008. Online at: *http://www.oecd.org/dataoecd/19/23/39693586.pdf.*

Peris-Lopez, P. et al. (2006) *RFID Systems: A Survey on Security Threats and Proposed Solutions*. December 2008. Online at: *http://www.springerlink.com/content/n5w675218t28h156/fulltext.pdf.*

Porter, M.E. and Millar, V.E. (1985) 'How information gives you competitive advantage,' *Harvard Business Review*, July–August: 149–60. December 2008. Online at: *http://zaphod.mindlab.umd.edu/docSeminar/pdfs/Porter85.pdf.*

Prabhu, B.S. et al. (2006) 'WinRFID – a middleware for the enablement of radio frequency identification (RFID) based applications'. In *Mobile, Wireless, and Sensor Networks – Technology, Applications, and Future Directions*, edited by R. Shorey et al., John Wiley.

R. Moroz Ltd. (2004) *Understanding Radio Frequency Identification (RFID) (Passive RFID)*. December 2008. Online at: *http://www.rmoroz.com/rfid.html.*

Ranganathan, S.R. (1957) *Five Laws of Library Science*. Madras: Madras Library Association.

Reding, V. (2006) *The RFID Revolution: Challenges and Options for Action*. December 2009. Online at: *http://*

europa.eu/rapid/pressReleasesAction.do?reference=SPEECH/06/162&format=PDF&aged=0&language=EN&guiLanguage=en.

RFID4u. *Library Automation using RFID*. December 2008. Online at: *http://www.rfid4u.com/downloads/Library%20Automation%20Using%20RFID.pdf.*

RNCOS (2009) *Global RFID Market Analysis till 2010: Summary*. November 2009. Online at: *http://www.rncos.com/Report/IM576.htm.*

Rogers, E.M. (2003). *Diffusion of Innovations*. New York: The Free Press.

Sabbaghi, A. and Vaidyanathan, G. (2008). 'Effectiveness and efficiency of RFID technology in supply chain management: strategic values and challenges,' *Journal of Theoretical and Applied Electronic Commerce Research*, 3(2): 71–81.

Schuyler, M. (2004) 'RFID: helpmate or conspiracy?,' *Computers in libraries*, 24(1): 22–4.

Shutzberg, L. (2004) *A Primer on RFID Technology and Processes: an Overview of Key Technology and Business Concepts that play into RFID Deployment*. December 2008. Online at: *http://www.informationweek.com/shared/printableArticle.jhtml?articleID=52500208.*

Smart, L. (2004) *Making Sense of RFID*. December 2008. Online at: *http://www.libraryjournal.com/article/CA456770.*

Smart, L.J. (2005) 'Considering RFID: benefits, limitations, and best practices,' *College and Research Library News*, 66(1). December 2008. Online at: *http://www.ala.org/ala/mgrps/divs/acrl/publications/crlnews/2005/jan/rfid.cfm.*

Smith, R. (2004) *RFID: A Brief Technology Analysis*. December 2008. Online at: *http://www.idspackaging.co.in/Common/Paper/Paper_253/RFID_A%20Brief%20Technology%20Analysis.htm.*

STLogiTrack. Online at: *http://www.stlogitrack.com.*

Stockman, H. (1948) 'Communication by means of reflected power,' *Proceedings of the Institute of Radio Engineers*, 1196–204.

TAGSYS. Online at: *http://www.tagsysrfid.com.*

Tellkamp, C. (2006) *The Impact of Auto-ID Technology Process Performance – RFID in the FMCG Supply Chain.* December 2008. Online at: *http://www.autoidlabs.org/uploads/media/AUTOIDLABS_WP_BIZAPP_025.pdf.*

Venkatesh, V. et al. (2003) 'User acceptance of information technology: toward a unified view,' *MIS Quarterly*, 27(3): 425–78.

Viehland, D. and Wong, A. (2007) 'Future of radio frequency identification,' *Journal of Theoretical and Applied Electronic Commerce Research*, 2(2): 74–81.

Want, R. (2004) 'The magic of RFID,' *Queue*, 2(7): 40–8.

Ward, M. and Kranenburg, R. (2006) *RFID: Frequency, Standards, Adoption and Innovation.* December 2008. Online at: *http://www.rfidconsultation.eu/docs/ficheiros/TSW0602.pdf.*

Weber, D. (2008) *Bibliotheca RFID.* December 2008. Online at: *http://www.gkri.hr/docs/gkriHR/documents/116/1.0/Original.pdf.*

Weinstein, R. (2005) 'RFID: a technical overview and its applications to the enterprise,' *IT Professional*, 7(3): 27–33.

Wu, N.C. et al. (2006) 'Challenges to global RFID adoption', *Technovation*, 26: 1317–23.

Wyld, D.C. (2005) *RFID: The Right Frequency for Government.* December 2008. Online at: *http://www.businessofgovernment.org/pdfs/WyldReport4.pdf.*

Wyld, D.C. (2006) 'RFID 101: the next big thing for management,' *Management Research News*, 29(4): 154–73.

Xang, X-D. et al. (2006) 'Review of RFID applications in global postal and courier services,' *Journal of China Universities of Posts and Telecommunications*, 13(4), 106–10.

Yu, S.-C. (2007) 'RFID implementation and benefits in libraries,' *Electronic Library*, 25(1): 54–64.

Index